Methods and Techniques in Deep Learning

IEEE Press
445 Hoes Lane
Piscataway, NJ 08854

IEEE Press Editorial Board
Sarah Spurgeon, *Editor in Chief*

Jón Atli Benediktsson   Andreas Molisch   Diomidis Spinellis
Anjan Bose   Saeid Nahavandi   Ahmet Murat Tekalp
Adam Drobot   Jeffrey Reed
Peter (Yong) Lian   Thomas Robertazzi

Methods and Techniques in Deep Learning

Advancements in mmWave Radar Solutions

Avik Santra
Souvik Hazra
Lorenzo Servadei
Thomas Stadelmayer
Michael Stephan
Anand Dubey

Infineon Technologies, Munich, Germany

IEEE PRESS
WILEY

Published by John Wiley & Sons, Inc., Hoboken, New Jersey.
Published simultaneously in Canada.

For general information on our other products and services or for technical support, please contact our Customer Care Department within the United States at (800) 762-2974, outside the United States at (317) 572-3993 or fax (317) 572-4002.

Wiley also publishes its books in a variety of electronic formats. Some content that appears in print may not be available in electronic formats. For more information about Wiley products, visit our web site at www.wiley.com.

Library of Congress Cataloging-in-Publication Data
Names: Santra, Avik, author.
Title: Methods and techniques in deep learning : advancements in mmwave
    radar solutions / Avik Santra, Souvik Hazra, Lorenzo Servadei, Thomas
    Stadelmayer, Michael Stephan, Anand Dubey, Infineon Technologies,
    Munich, Germany.
Description: Hoboken, New Jersey : John Wiley & Sons, Inc., [2023] |
    Includes bibliographical references and index.
Identifiers: LCCN 2022036520 (print) | LCCN 2022036521 (ebook) | ISBN
    9781119910657 (hardback) | ISBN 9781119910664 (adobe pdf) | ISBN
    9781119910671 (epub)
Subjects: LCSH: Millimeter wave radar–Data processing. | Radar
    targets–Identification–Data processing. | Radar receiving
    apparatus–Data processing. | Deep learning (Machine learning)
Classification: LCC TK6592.M55 S26 2023 (print) | LCC TK6592.M55 (ebook)
    | DDC 621.38480285–dc23/eng/20220929
LC record available at https://lccn.loc.gov/2022036520
LC ebook record available at https://lccn.loc.gov/2022036521

Cover Design: Wiley
Cover Image: © Yurchanka Siarhei/Shutterstock

Set in 9.5/12.5pt STIXTwoText by Straive, Chennai, India

This book is dedicated to our respective families.

Contents

About the Authors

Avik Santra

Avik Santra received his B.S. degree in electronics and communications engineering from West Bengal University of Technology. He then received his M.S. degree in signal processing with first-class distinction from Indian Institute of Science and a Ph.D. degree in electrical, electronics, and informatics from the FAU University of Erlangen, Germany. He is currently heading the advanced AI team responsible for developing signal processing and machine learning algorithms and system solutions for radars and depth sensors at Infineon, Germany. Earlier in his career, he worked as a system design engineer for LTE chipsets at Broadcom Communications developing and implementing calibration algorithms for LTE chipsets. Subsequently, he has worked as a research engineer developing system concepts and representative demonstrators of next-generation long-range radars and data analytics at Airbus. He has received several spot awards for project excellence in multiple forums. He has been an invited speaker at various conferences and workshops, as well as delivered several tutorials on deep learning and signal processing topics. He is a reviewer at various IEEE and Elsevier journals and is a recipient of several outstanding reviewer awards. He has been lead guest editor at *IEEE Sensors Journal* and associate editor at Elsevier Machine Learning with Applications. He is the coauthor of the book *Deep Learning Applications of Short-Range Radars*, published by Artech House in September 2020. He has filed over 70 US/EU patents and published over 55 research papers related to deep learning and signal processing topics. He is a senior member of IEEE.

Souvik Hazra

Souvik Hazra received his B. Tech degree in electrical engineering from KIIT University in 2017 and then received his MS degree in data science and engineering from EURECOM and IMT, France, in 2019. He is currently working as a senior staff machine learning engineer at Infineon Technologies AG, Munich, where he is responsible for the overall development of machine learning and signal processing solutions for radars and microphones. Earlier in his career, he has worked as a research intern at Airbus and CCAF, University of Cambridge, on various deep learning topics. He has been invited as a speaker at various summits and has been a reviewer at various IEEE journals and conferences. He is the coauthor of the book *Deep Learning Applications of Short-Range Radars*, published by Artech House in September 2020. Besides his full-time job at Infineon, he is pursuing his PhD degree at Friedrich-Alexander-University (FAU), Erlangen.

Lorenzo Servadei

Dr. Lorenzo Servadei is a senior staff machine learning engineer at Infineon Technologies AG. His main interests are methods of reinforcement learning applied to quantum computing design, signal processing, and design automation of microchips. He obtained a PhD in computer science from a collaboration between Infineon Technologies AG and Johannes Kepler University of Linz.

His PhD focused on the use of methods of reinforcement learning for hardware design optimization. To this end, he researched and approached methods of combinatorial optimization for the improvement of power, performance, and area (PPA) on digital hardware. In particular, he developed combinatorial reinforcement learning (RL) algorithms that gradually improve the positioning and connection of subcomponents within the hardware schematic. During his PhD, he collaborated and published two journal papers with Professor Hochreiter, inventor of the Long-Short Term Memory (LSTM) networks. Additionally, he spent several months at Duke University, working on machine learning contributions on Hardware Security. Dr. Lorenzo Servadei has also served as a machine learning trainer for Infineon Technologies AG, helping to grow the artificial intelligence community within the company in different sites around the world. He is currently an IEEE member, a senior lecturer, Habilitand, and a post-doc in the Department of Electrical and Computer Engineering at the Technical University of Munich.

Thomas Stadelmayer

Thomas Stadelmayer was born in Regensburg, Bavaria, Germany, in 1994. He studied computational engineering at Friedrich-Alexander University (FAU) Erlangen-Nuremberg and graduated with a bachelor's degree in 2015 and a master's degree in 2018. He then worked as a research assistant in the Circuits, Systems and Hardware Test (CST) research group at the Institute of Electrical Engineering, FAU Erlangen-Nuremberg. His research interests include digital signal processing and machine learning for short-range and indoor radar applications. During his research, he worked in close collaboration with Infineon Technologies on various radar applications based on machine learning, such as hand gesture recognition and person localization. He is particularly interested in combining classical signal processing and machine learning to obtain more interpretable neural networks. He is also interested in deep metric learning for detecting outliers or unknown motion to make applications more robust in real-world environments. He has contributed to his research area with scientific publications and several patent applications. He joined Infineon Technologies in February 2022 in the Advanced Artificial Intelligence group. His task is to improve current signal processing-based radar algorithms using artificial intelligence to overcome application limitations while also exploring new applications enabled by artificial intelligence for short-range radars. He also builds proof-of-concept demonstrators and works closely with academic partners. Besides his work at Infineon, he is pursuing his PhD degree.

Michael Stephan

Michael Stephan was born in Forchheim, Bavaria, Germany, in 1995. He received his bachelor's degree in electrical engineering and master's degree in advanced signal processing and communications engineering from the Friedrich-Alexander-University Erlangen-Nuremberg in 2017 and 2019, respectively. During his studies, he was a visiting scientist at Nokia Bell Labs, Holmdel, New Jersey, USA, where he looked into the RF-chains for Hybrid MIMO Precoders. He wrote his master thesis at the Poly-Grames Research Center in Montréal, Canada, about algorithmically reducing the effect of coupling on the angle of arrival estimation performance of MIMO FMCW radar and also completed an internship at Infineon Technologies AG in Linz, Austria, marking his first contact with deep learning for indoor target localization using FMCW radar sensors. He is currently pursuing his PhD degree with the Friedrich-Alexander-University Erlangen-Nuremberg at the Institute for Electronics Engineering in cooperation with Infineon Technologies AG, Neubiberg, Germany. Coming from a signal processing perspective, his current research focuses on various deep learning topics with application in mmWave radar signal processing for indoor localization and tracking in real-world environments. He has written numerous publications and filed multiple patents on deep learning applications for radar processing. Recently, his research focuses on explicitly using traditional signal processing knowledge during the neural network training process to achieve better generalization and performance in the low-data regime.

Anand Dubey

Anand Dubey was born in Mirzapur, Uttar Pradesh, India, in 1990. He studied electronics and communication engineering at Jaypee Institute of Information Technology University (JIITU) for his bachelor's degree in 2012 and automotive software engineering at Technical University of Chemnitz for his master's degree in 2018. Later, he worked as a research assistant in the Circuits, Systems and Hardware Test (CST) research group at the Institute of Electrical Engineering, FAU Erlangen-Nuremberg. His research interests include digital signal processing and machine learning for automotive radar applications. During his research, he worked on an application to detect and classify pedestrians and cyclists using their motion and spatial signatures. He is particularly interested in combining statistical signal processing and Bayesian machine learning to obtain more interpretable and reliable neural networks. He is also interested in the domain of geometric learning where data are sparse and correlated. He has contributed to his research area with several scientific publications. He joined Infineon Technologies in January 2022, and his task is to investigate and propose novel signal processing pipelines for speech enhancement using Bayesian machine learning algorithms. He is also exploring areas of tiny machine learning algorithms for low-powered, microcontroller units.

Preface

Radar has evolved from a complex, high-end military technology into a relatively simple, low-end solution penetrating industrial, automotive, and consumer market segments. This rapid evolution has been driven by the advancements in silicon and the use of deep learning algorithms to utilize the full potential of sensor data. The use of radar sensors has grown multifold penetrating automotive, industrial, and consumer markets, offering a plethora of applications. The advent of deep learning has transformed many fields and resulted in state-of-the-art solutions in computer vision, natural language processing, speech processing, etc. However, the application of deep learning algorithms to radars is still by and large at its nascent stage. This book attempts to present the theoretical concepts behind several advanced deep learning concepts and highlight how such techniques enable such applications, which were not otherwise possible.

This book presents cutting-edge artificial intelligence (AI)-based processing using advanced deep learning to a short-range radar. AI is the hottest topic in all industrial sectors and has led to disruptions across all fields, such as computer vision, natural language processing, speech processing, medical imaging, etc. However, the application of AI to radars is relatively new and unexplored. We in this book present the cutting-edge deep learning processing that we worked and are working on at Infineon Technologies. This book covers how advanced deep learning concepts are being used to enable applications ranging from industrial sector, consumer space, to emerging automotive industries. This book presents examples of several human–machine interface applications such as gesture recognition and sensing, human activity classification, people counting, people localization, and tracking along with automotive target detections, localization, and classification.

Chapter 1 introduces the fundamentals of deep learning, its evolution over time, and the different facets that make deep learning so powerful. This chapter introduces various components of conventional convolutional neural networks, recurrent neural network, and fully connected layers in relation to various tasks such as classification, localization, segmentation, or translation. Chapter 2 presents deep metric learning with an intensive overview of the state-of-the-art algorithms and how open-set classification tasks are handled using metric learning. Then, a short-range radar application that aims to classify among a set of predefined hand gestures amid random unknown motions is presented.

Chapter 3 introduces deep parametric learning, where the preprocessing pipeline can be integrated into a deep neural network and made data driven, thus enhancing the performance to be task specific as well as making the architecture compact. Chapter 4 introduces deep reinforcement learning, where the learning algorithm depends on the sum of rewards produced by policy interacting with an environment. We review the basics of deep reinforcement learning and then present the overview of different typologies of deep reinforcement learning algorithms. We present the efficacies of deep parametric learning with activity classification application, and for reinforcement learning, we present how it helps to update the parameters of a tracker adaptively as a function on the target dynamics.

Chapter 5 introduces cross-modal learning algorithms by giving an overview of the state-of-the-art approaches, and then, we present two approaches of cross-modal learning to improve radar-based people counting solutions in comparison with unimodal learning approaches. In Chapter 6, we present signal processing-led learning that gives an overview over different model-based approaches to incorporate expert knowledge in deep learning methodologies. We present the advantages of signal processing-driven deep learning with respect to radar-based target detection and segmentation use case.

Chapter 7 presents domain adaptation wherein the model is trained on a source data distribution and then deployed for a different target data distribution. Transfer learning and fine-tuning are subsets of domain adaptation, and here, we present the overview of the existing techniques and introduce them to specific applications of human activity classification. Chapter 8 presents Bayesian deep learning, introducing an overview on the history of learning theory for deterministic and Bayesian neural networks followed by understanding on different elemental blocks required to formulate Bayesian deep learning and then a practical application demonstrating the efficacies of Bayesian deep learning for an automotive radar. Chapter 9 introduces geometric deep learning, starting with the overview followed by the need to capture and learn underlying patterns in a complex non-Euclidean data structure. Subsequently, practical application is demonstrated using automotive radar point clouds for automotive target classification and for long-range gesture sensing.

This book is intended for graduate students, academic researchers, and industry practitioners working with deep learning who strive to apply deep learning techniques to mmWave radars or depth sensors. This book is written keeping beginners to advanced researchers in mind and assumes sufficient knowledge of linear algebra and engineering mathematics. Each chapter has end questions to assess the understanding of the reader. This book covers the theoretical foundation of each deep learning algorithm or paradigm and also presents the adaptations of such algorithms to a specific mmWave radar application. This book covers advanced concepts such as deep metric learning, parametric learning, reinforcement learning, reinforcement learning, cross-learning, signal processing-led architectures, domain adaptation, and geometric deep learning. While each chapter is independent of the other, it is suggested that an early researcher reads the first introductory chapter introducing basic radar signal processing and deep learning before reading the specific deep learning chapter.

The authors would like to express their heartfelt gratitude to their PhD supervisors Prof. Robert Weigel and Prof. Robert Wille for their constant guidance and support. We look at them with great respect for their profound knowledge and experience, their unparalleled teaching and problem-solving skills, and their relentless pursuit for perfection, which has something we try to emulate all the time and in this book.

We are thankful to our department head, Gerhard Martin, for being extremely supportive and encouraging us all the time to give our best. We would also like to greatly thank Dr. Christian Mandl who has been a lighthouse of inspiration for us, guiding us with his accurate understanding of technical concepts along with his leadership skills. We would also like to thank Dr. Ashutosh Pandey for the technical guidance and unparallel knowledge and his relentless strive for excellence.

The authors would also like to thank their editors and reviewers for his encouragements, reviews, and suggestions to improve this book.

Acronyms

BDL	Bayesian deep learning
BIC	Bayesian information criterion
BNN	Bayesian neural networks
CFEL	Complex frequency extraction layer
CV	Computer vision
CNN	Convolutional neural network
CM	Confusion matrix
CFAR	Constant false alarm rate
DDPG	Deep deterministic policy gradient
DRL	Deep reinforcement learning
DBSCAN	Density-based spatial clustering of applications with noise
DSP	Digital signal processor
DoA	Direction of arrival
DA	Domain adaptation
d-SNE	Domain adaptation using stochastic neighborhood embedding
DGCNN	Dynamic graph CNN
EKF	Extended Kalman filter
FADA	Few-shot adversarial domain adaptation
FMCW	Frequency-modulated continuous-wave
GMM	Gaussian mixture model
GP	Gaussian process
GRV	Gaussian random variable
GAN	Generative adversarial network
GRL	Gradient reversal layer
GAE	Graph autoencoder
GNN	Graph neural network
IID	Independent and identically distributed
ISTA	Iterative shrinkage-thresholding algorithm
K-NN	k-nearest neighbor

LAR	Label-aware ranked
LIDAR	Laser imaging detection and ranging
lse	log-sum-exp
LSTM	Long short-term memory
ML	Machine learning
MDD	Margin disparity discrepancy
MCMC	Markov chain Monte Carlo
MDD	Maximum mean discrepancy
MH	Metropolis–hastings algorithm
mLSTM	multifrequency long short-term memory
mWDN	multilevel wavelet decomposition network
MCB	Multimodal compact bilinear
MMDL	Multimodal deep learning
NN	Neural network
NMS	Nonmaxima suppression
OKS	Object keypoint similarity
OS-CFAR	Ordered statistics CFAR
RAI	Range-angle image
RDI	Range-doppler image
RGNN	Recurrent graph NN
RNN	Recurrent neural network
RL	Reinforcement learning
RCF	Residual classification
SPKF	Sigma-point Kalman filter
STGNN	Spatial-temporal graph neural networks
SVM	Support vector machine
TL	Transfer learning
TRPO	Trust region policy optimization
VAE	Variational auto-encoder
VBGM	Variational Bayesian Gaussian mixture model
VI	Variational inference
VQA	Visual question answering
VRU	Vulnerable road users flow
UMAP	Uniform manifold approximation and projection
UKF	Unscented Kalman filter

1

Introduction to Radar Processing and Deep Learning

At the end of this chapter, reader will have understanding on

- How radar data cubes are processed to extract range, velocity, and angle of the multiple detected targets and are tracked over time.
- Different target representations used for radar target recognition.
- Introduction to deep learning architectures used for radar target recognition.

1.1 Basics of Radar Systems

Radar is an acronym that stands for radio detection and ranging. It is basically an electromagnetic system used to detect the presence of one or more targets of interest and estimate their range, angle, and velocity relative to the radar. Instead of just measuring the target's location and velocity, modern radars can predict the target given the reflected radar signals. The main objective of radar compared to infrared and optical sensors is to discover distant targets under difficult climate conditions and to determine their spatial location while tracking them over time with precision. The general working principle and signal processing fundamental details are explained in the following sections.

1.1.1 Fundamentals

The radar system generally consists of a transmitter that produces an electromagnetic signal, which is radiated into space by the transmit antenna. When this signal strikes an object, it gets reflected or re-radiated in many directions. This reflected echo signal is received by the receive antenna, which delivers it to the receiver circuitry, where it is processed to detect the target and also localize it over time along with certain characteristics of the target. A simplified version of a typical continuous wave radar front-end with the most important building blocks can be

Methods and Techniques in Deep Learning: Advancements in mmWave Radar Solutions, First Edition.
Avik Santra, Souvik Hazra, Lorenzo Servadei, Thomas Stadelmayer, Michael Stephan, and Anand Dubey.
© 2023 The Institute of Electrical and Electronics Engineers, Inc. Published 2023 by John Wiley & Sons, Inc.

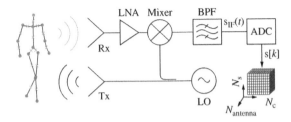

Figure 1.1 Block diagram of the continuous wave radar front-end and its receive chain including the mixer, band pass filter, and analog-to-digital converter. The digitized samples $s_{IF}(t)$ are stored into a data matrix $s[k, l]$. The radar in this case is sensing a human target in the field of view.

seen in Figure 1.1. The chosen waveform is generated by a local oscillator (LO) and transmitted via the transmit (Tx) antenna. The receive (Rx) antenna then captures the incoming signal reflections from the target at a distance. After amplifying the received signal, it is mixed with the original transmitted waveform and is passed through subsequent analog bandpass filtering (BPF). This removes any high-frequency components that could cause aliasing as well as low-frequency components from direct coupling of the LO signal into the receiver. After mixing and filtering the signal that has been shifted to an intermediate frequency (IF), and it is referred as $s_{IF}(t)$. The IF bandwidth B_{IF} is determined by the upper cut-off frequency of the bandpass filter, which is typically in the order of tens of kHz to few MHz.

1.1.2 Signal Modulation

To detect and differentiate multiple targets along its range, relative velocity, and azimuth-elevation angle dimensions, linear frequency modulated continuous wave (FMCW) is used as the most standard sensing waveform [1]. Usually, consecutive identical chirps are transmitted within a frame with a predefined time spacing referred to as chirp repetition time. The received IF signal is arranged within a two dimensional matrix, and the intratime, i.e., within a chirp, is referred as fast-time, while the intertime, i.e., across chirps, is referred to as slow-time. If the target is static, the round trip delay in the received signal is manifested as a frequency offset along the fast-time dimension after down-mixing at the receiver. But if the target or the radar is not stationary, the received signal will have an additional frequency offset caused by the Doppler shift manifested across slow-time dimension.

Figure 1.2 shows the concept of a FMCW modulation in detail. The LO generates a chirp signal $s_{LO}(t)$ with starting frequency $f_{0,LO}$, bandwidth B_c, duration T_c, and resulting sweep rate $\mu_{LO} = B_c/T_c$. By taking advantage of time integral over

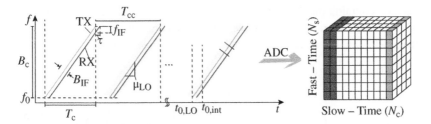

Figure 1.2 Illustration of typical modern radar sensors with several identical transmit chirps within a frame and the digitized IF samples $s_{IF}[k]$ are then stored chirpwise in a data matrix for coherent processing.

Tx frequency, instantaneous phase is calculated as shown in Eq. (1.1), where $\varphi_{0,LO}$ corresponds to the initial phase of the LO:

$$f_{Tx}(t) = f_{0,LO} + B_c/T_c t \tag{1.1a}$$

$$\phi_{Tx}(t) = 2\pi \int_0^{T_c} f_{Tx}(t)dt = 2\pi f_{0,LO}t + \pi \frac{B_c}{T_c}t^2 + \varphi_{0,LO} \tag{1.1b}$$

Assuming unity amplitude for a single chirp, $s_{LO}(t)$ can be formulated by

$$s_{Tx}(t) = \begin{cases} \cos\left(\phi_{Tx}(t)\right) & 0 \leq t \leq T_c \\ 0 & \text{else} \end{cases} \tag{1.2}$$

If this transmit signal gets reflected by some object, also referred to as target, the reflection will be received at the radar with a time delay τ, which is proportional to the target's distance to the radar. Additionally, signals of multiple reflections, like extended targets, are superimposed on each other at the receiver. For an arbitrary number of point targets N_{tgt} composing a spatially distributed target, the received signal $s_{Rx}(t)$ can thus be expressed as follows:

$$s_{Rx}(t) = \sum_{i=1}^{N_{tgt}} s_{Tx}(t - \tau_i(t)) + n \tag{1.3}$$

where n represents thermal receiver noise or clutter and $\tau_i(t)$ is the round trip delay to the ith target located at distance R_i and moving with a relative radial velocity of v_i. As a result, $\tau_i(t)$ can be described as $\frac{2R_i - 2v_i t}{c_0}$, where c_0 is the speed of light. For ease of notations, the noise term is dropped for all the following considerations. The received and amplified signal is mixed with the original transmitted signal ($s_{mix}(t) = s_{Tx}(t) \cdot s_{Rx}(t)$). As discussed before, both transmitted and received signal follows cosine waveform. Thus, the down-mixed signal $s_{mix}(t)$ can be transformed into two components using trigonometric formulation.

$$s_{mix}(t) = s_{Rx}(t) \cdot s_{Tx}(t) = \frac{1}{2}(\cos(\phi_{diff}(t)) + \cos(\phi_{sum}(t))) \tag{1.4}$$

here $\phi_{\text{diff}}(t)$ contains the difference of Tx and Rx signal frequencies and $\phi_{\text{sum}}(t)$ contains the sum frequencies respectively. The sum component is removed by the following BPF and the resulting IF signal $s_{\text{IF}}(t)$ is obtained as follows:

$$s_{\text{IF}}(t) = \frac{1}{2}\cos(\phi_{\text{diff}}(t))$$

$$\approx \frac{1}{2}\cos\left(2\pi\left(\underbrace{\frac{2f_0 R}{c_0}}_{\phi_0} + \overbrace{\left(\underbrace{\frac{2B_c R}{c_0 T_c}}_{f_R} - \underbrace{\frac{2f_0 v}{c_0}}_{f_D}\right)}^{f_b}t - \frac{2B_c v}{c_0 T_c}t^2\right)\right) \tag{1.5}$$

This shows that intermediate received signal contains both distance-dependent frequency f_R and also speed-dependent frequency shift f_D which are factors of modulation parameters. This includes chirp duration T_c, chirp repetition time T_{cc}, sweep frequency or bandwidth B_c, and number of chirps in a frame N_c as main configuration parameter for the design of a FMCW waveform. As a result, these parameters control the range and Doppler resolution, as presented in Eqs. (1.6) and (1.7), respectively. The maximum observable range and the maximum unambiguous Doppler is given in Eqs. (1.8) and (1.9).

$$\Delta f_R = \frac{2B_c R}{c_0 T_c} \tag{1.6a}$$

$$\Delta R_{\min} = 1.21\frac{c_0}{2B_c} \tag{1.6b}$$

$$\Delta f_v = \frac{2f_0 v}{c_0} \tag{1.7a}$$

$$\Delta v_{\min} = 1.21\frac{c_0}{2f_0 N_c T_{cc}} \tag{1.7b}$$

$$R_{\max} = \frac{f_s c_0 T_c}{2B_c} \tag{1.8}$$

$$V_{\max} = \frac{c_0}{2T_{cc} f_0} \tag{1.9}$$

1.2 FMCW Signal Processing

The IF signal, obtained from Eq. (1.5), is then digitized by an analog-to-digital converter with sampling period T_s at the discrete time instants kT_s, where $k \in [0, ..., N_s - 1]$. Consequently, the discrete time signal $s[k]$ contains N_s samples per chirp. Typically, modern short-range radar sensors rapidly transmit several identical chirps in a so-called chirp sequence modulation. The digitized IF samples $s[k]$ are then stored chirp wise in a data matrix $s[k, l]$ for coherent processing. Figure 1.3 summarizes the FMCW signal processing multistage pipeline, where

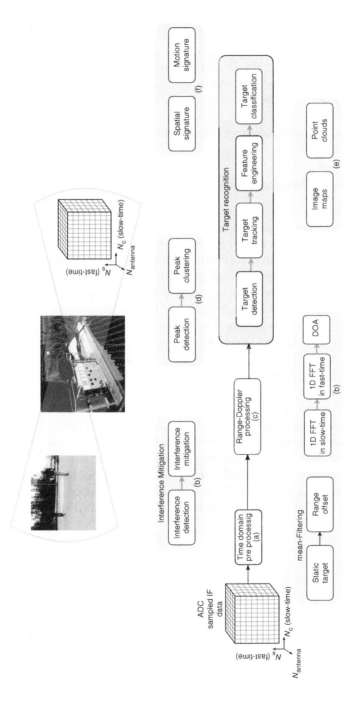

Figure 1.3 Summary of FMCW signal processing pipeline including both pre- and postprocessing over $s[k, l]$ chirp matrix for target detection, tracking, and classification.

$s[k, l]$ is first preprocessed in time-domain for removal of spectral leakage or static targets followed by interference mitigation. Later, the preprocessed $s[k, l]$ is transformed to frequency domain for target detection. Once a target is detected, then the measurement is fed into a tracking algorithm for temporal smoothening. At the end, the tracked target's features are extracted using their motion or spatial signature in the form of images or point-clouds, respectively, which are used for target recognition.

The frequency domain analysis for range-Doppler processing is explained in the following section.

1.2.1 Frequency-Domain Analysis

As indicated by Eqs. (1.6) and (1.7) both range and radial velocity information of the target are functions of frequency shifts in the received signal. As a result, frequency-domain analysis is used to determine respective target's parameters instead of time-domain analysis. In contrast to time-domain signals where signal changes over time (amplitude or power) can be observed, frequency-domain analysis reveals how much of the signal lies within each given frequency band over a range of frequencies, which also include change in phase information. The most common frequency-domain transform methods are Fourier transform, short-time Fourier transform (STFT) and wavelet transforms. All three transforms are inner products of a family of basis functions with a time-domain signal. The parameterization and the basis functions determine the properties of the transforms.

Before delving into details, Figure 1.4 illustrates all the three transforms pictorially. While the classical technique to represent time signals in the frequency domain by calculating discrete Fourier transformation (DFT), it fails to detect time variant frequency effects, which are important for extended targets. As an alternative, DFT is modified by shortening the time window for each DFT and leads to short-time Fourier transformation (STFT), which improves resolution in time but at the cost of lower accuracy in the frequency domain, as seen in Figure 1.4. While the DFT has no temporal resolution and STFT have fixed

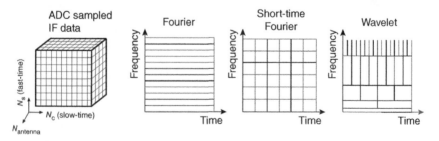

Figure 1.4 Pictorial representation of all three transforms, i.e., Fourier, STFT, and wavelets over analog-to-digital conversion (ADC) sampled chirp sequence data.

resolution for complete time-frequency, the wavelet transformation can adapt both the time and frequency dimensions and result into a high-frequency resolution at low frequencies while maintaining good time localization at high frequency.

1.2.1.1 Discrete Fourier Transform

While the Fourier series is used for oscillating or repetitive signal, Fourier transform is used for nonrepetitive signals. Thus, Fourier transform can be formulated as a special case of Fourier series when the time period $T \to \infty$. For the standard Fourier transform, the basis functions are simply the complex sinusoidal oscillations

$$b_\omega(t) = e^{j\omega t} \tag{1.10}$$

where t is the time axis of the signal and ω is the single frequency parameter that determines the basis function in the family. There is one basis function for every ω. The Fourier transform of the signal $x(t)$ is then simply the inner product written as an integral

$$x(\omega) = \mathcal{F}\{x(t)\}(\omega) = \langle b(\omega, t), x(t) \rangle = \int_{-\infty}^{\infty} e^{-j\omega \tau} x(\tau) d\tau \tag{1.11}$$

The negative sign in the exponential comes from the complex conjugation in the general inner product definition:

$$\langle a(t), b(t) \rangle = \int_{-\infty}^{\infty} a(\tau)^* b(\tau) d\tau \tag{1.12}$$

Since we are processing down-sampled raw radar analog-to-digital conversion (ADC) data matrix, all signals are considered as discrete time and valued (unlike continuous signals). Thus, discrete time Fourier transform (DTFT) is performed as follows:

$$X(\omega) = \sum_{n=-\infty}^{\infty} x(n) e^{-j\omega n} \tag{1.13}$$

where $x(n)$ is discrete time-value signal and $X(\omega)$ is continuous spectrum. Consequently, as $X(\omega)$ is of continuous nature, the DTFT cannot be processed directly on a digital machine. Therefore, a discrete spectrum response is required, which is usually done by using the discrete Fourier transform (DFT) and denoted by $X(k)$. This is done by sampling over the spectrum of the DTFT. The sampling frequency of DFT is defined as $\omega_k = 2\pi k/N$. Here, N corresponds to the total number of samples retrieved from the DTFT:

$$X\left(\frac{2\pi k}{N}\right) = \sum_{n=0}^{N-1} x(n) e^{-j2\pi kn/N} \tag{1.14}$$

$$X(k) = \sum_{n=0}^{N-1} x(n) e^{-j2\pi kn/N} \tag{1.15}$$

The above equation shows two fundamental mathematical operations, which are carried out for every sample of the input signal – multiplication and addition. The DFT is an iterative operation and requires a high-computational effort. The fast-Fourier transform (FFT) is an algorithm to compute the DFT efficiently. This is usually done using the Cooley–Tukey algorithm; however, there exist many other algorithms.

The FFT operation is applied over down-sampled raw radar ADC data matrix stored along the chirp, namely fast-time dimension, with consecutive chirps stored along the columns, referred to as slow-time dimension. Prior to the FFT operation, as seen in Figure 1.3, time-domain preprocessing can be done as an optional step. This helps to remove Tx–Rx leakage and clutter noise from down-sampled raw radar ADC data matrix. This is done by mean subtraction across fast-time and slow-time, respectively. This process is commonly known as moving target indicator (MTI) processing. Furthermore, an optional interference detection and mitigation method could also be applied to reduce the effect of interference and noise. By calculating the two-dimensional fast Fourier transform (FFT) on this data matrix, fast-time is converted to range-frequency and slow-time to Doppler frequency. This operation yields the spectrum S_{RD}, with indexed dimensions as range-frequency f_r and Doppler-frequency f_v.

In principle, FFT assumes that the signal contains a continuous spectrum that is one period of a periodic signal. However, the measured raw data matrix $s[k, l]$ may not contain integer number of periods. Therefore, the definiteness of the $s[k, l]$ may result in a discontinuity at the endpoints of the waveform in comparison to the original continuous-time signal and could introduce sharp transition changes into the consecutive measured signals. These artificial discontinuities lead to the additional high-frequency components not present in the original signal. This phenomenon is called spectral leakage where energy at one frequency leaks into other frequencies. This causes the sharp frequency spectrum to spread into wider signals and leads to ambiguity.

These effects are minimized using a technique called windowing. The windowing function reduces the amplitude of the discontinuities at the boundaries of each finite sequence acquired by the ADC. Windowing function consists of amplitude envelope that is multiplied elementwise with the original ADC matrix. The characteristic of windowing function is such that it varies smoothly and gradually toward zero at the edges. This makes the endpoints of the measurement $s[k, l]$ similar and therefore results in a continuous waveform without sharp transitions. In addition to it, in general, zero-padding along either of dimension of the data matrix $s[k, l]$ is done with a factor of power of 2. This interpolates the coarse spectrum to become more smooth but does not reveal extra information from the spectrum. To improve the resolution of the spectrum, length of the recorded signal needs to be increased. Additionally, it can also be interpreted as windowing, which is time-domain multiplication of rectangular function with the original signal.

1.2.1.2 Short-Time Fourier Transform (STFT)

Since FFT is performed over complete ADC sampled chirp matrix $s[k, l]$, it averages out signal frequency components. This approach is good for localization and detection of reflection from targets but fails to extract both spatial or motion information (commonly termed as signatures) for an extended targets like humans.[1] As a result, FFT is modified by adding time dimension to the base function $(b_{(\omega,t_0)}(t))$ parameters by multiplying the infinitely long complex exponential with a window to localize it. This transform is known as STFT, whose base functions are then

$$b_{(\omega,t_0)}(t) = w\left(t - t_0\right)\exp(i\omega t) \tag{1.16}$$

where $w(t)$ is the window functions that vanish outside some interval and $\left(\omega, t_0\right)$ are the time-frequency coordinates of the base function in the family. The inner product is formulated as follows:

$$S\{s(t)\}\left(\omega, t_0\right) = \int_{-\infty}^{\infty} w(\tau)^* \exp(-i\omega\tau)s(\tau)d\tau \tag{1.17}$$

The advantage of STFT is that it can capture frequency information of target's signature over time, i.e., range signature or micro-Doppler signatures. This information can be treated as unique signature for target recognition or its attribute recognition. However, it is challenging to find a trade-off between time and frequency resolution while calculating the STFT. This is determined by the choice of the window function and sampling frequency.

1.2.1.3 Wavelets

In contrast to DFT and STFT, the wavelet transform adapts the window size to the frequency with the constant bandwidth constraint. This is designed in a scale invariant approach that doesn't even need the complex modulation basis function. The working principle of wavelet transforms can be understood with a generic base function that is localized and oscillates with zero mean, i.e., the integral over the complete space is zero. This basis function is referred as mother wavelet. The advantage of such a basis function (wavelet) is that both localization (time) and oscillation (frequency) resolution trade-off can be reduced, which is a constraint in STFT. Therefore, the family of base functions can be summarized as

$$b_{(\sigma,t_0)}(t) = w\left(\frac{t - t_0}{\sigma}\right) \tag{1.18}$$

where w is the mother wavelet and σ the scale parameter. Thus, the inner product becomes

$$\mathcal{W}\{s(t)\}\left(\sigma, t_0\right) = \int_{-\infty}^{\infty} w\left(\frac{t - t_0}{\sigma}\right)^* s(t) \tag{1.19}$$

---

1 With the availability of high-resolution radar, human targets are treated as doubly extended targets due to their multiple spatial reflections along range and along Doppler referred to as micro-Doppler components.

1.3   Target Detection and Clustering

As illustrated in Figure 1.3, irrespective of the target representation, i.e., images or point clouds, the most standard target detection has two stages, involving the detection followed by clustering. A simple approach to target detection is peak detection by determining if the sensed bin has higher amplitude than its neighboring bins.

Alternately, constant false alarm rate (CFAR) detector is used for detection of the targets. CFAR detector calculates detection probability for each bin by estimating varying noise power from neighboring cells as shown in Eq. (1.20). Here, T refers to detection threshold, α is scaling factor, P_n is estimated noise power, N counts the total number of neighboring reference cells and P_{fa} is the CFAR. Equation (1.20) represents cell averaging (CA)-CFAR. The drawback of CA-CFAR is that it occludes the weak targets near to strong target by having higher noise threshold and thus masking it out. As an alternative, ordered statistics (OS)-CFAR is being used for detection. The kth ordered data is selected as threshold instead of averaging over all reference cells.

$$T = \alpha P_n \tag{1.20a}$$

$$P_n = \frac{1}{N} \sum_{m=1}^{N} x_m \tag{1.20b}$$

$$\alpha = N \left(P_{fa}^{-1/N} - 1 \right) \tag{1.20c}$$

In contrast to rigid targets like car, truck, human as target of interest contains micro-motions which results in different velocity component in the received signal, commonly known as micro-Doppler components [2, 3]. This results to a spread of detected targets across Doppler dimension. Also, with the use of higher sweep bandwidth, the range resolution of the radar lies in the order of few centimeters. As a result, the reflection from target are not received as point target reflection but are spread across multiple range bins. These targets are also known as range-Doppler extended targets or doubly spread targets. As a result, all the detection from target needs to be clustered into one and thus necessitates clustering algorithm as the second stage. This also helps in reducing the computational complexity for the target-tracking algorithm, which after clustering tracks a single target parameter instead of tracking nonclustered group of target parameters. Density-based spatial clustering of applications with noise (DBSCAN) is used [4] as the state-of-the-art algorithm. Unlike most algorithms, DBSCAN runs clustering in one pass without having prior knowledge on number of clusters and is stable to the outliers (noise). The input hyperparameters required for DBSCAN are a minimum number of points M and minimum distance d between neighboring points to be part of cluster [5]. Given a set of target detections from same and multiple targets in the 2D space, DBSCAN groups detections that are closely packed together, while at the same time removing as outliers detections that lie alone in low-density regions. To do this, DBSCAN classifies each point as either core point, edge point, or noise. A point is defined as core point if it has at least $M - 1$ neighbors, i.e., points within the distance d. An edge point has less than $M - 1$ neighbors, but at least one of its neighbors is a core point. All points that have less than $M - 1$ neighbors and no core point as a neighbor do not belong to any cluster and will be classified as outlier and ignored. The two-stage approach has limitations in both the stages. While OS-CFAR fails to detect target in the case of clutter, multipath reflection, or interference for a fixed P_{fa}, DBSCAN fails to make cluster for sparse range-Doppler images (RDIs) and also is very sensitive to its hyperparameters. As a result, it may either lead to false detection, miss detections of real targets or target splits leading to multiple targets (when in reality it is only one target) in RDIs. However, the recent advancement in deep neural networks (DNNs) and its application in target segmentation makes DNNs an ideal algorithm for this problem. Unlike fixed rule-based methods, DNNs are capable of learning from low-level to high-level representations. The problem of two-stage target detection is treated as binary image segmentation problem in literature where target's cluster is considered as foreground and remaining information in range-Doppler map as background, as illustrated in Figure 1.5b. In [6, 7], authors have successfully demonstrated single-stage target detection on RDIs while suppressing the effect of scatterings from extended targets, multipath reflections and ghost targets. Further, in [8] single-stage target detection is introduced that preserves the instance of each target to identify them uniquely in case of a partial occlusion or overlapping of multiple targets. Additionally, [8] gives an alternative approach to combine interference mitigation and target detection by selectively looking into regions of interest.

However, the performance of target detection using image-based maps (most commonly range-Doppler and range-angle) rely strongly on the target's spread distribution pattern, which further depends on the target's aspect angle. However, by taking advantage of strong feature points from the target distribution and learning a relation between these feature points and values, the target detection can be made robust. Thus, it makes sense to transform the radar images to radar point-clouds, as shown in Figure 1.5c. Most approaches from literature propose to process radar point-clouds using certain representation transformations in order to use convolutional neural networks (CNNs), e.g., radar data is transformed into a grid map representation. However, [9–11] proposed a novel neural network, PointNet, that makes it possible to directly process point clouds. By making the representation invariant to transformation, PointNet additionally treats neighboring points to capture local structure and is invariant to structure of input point clouds. Although, radar data are very sparse compared to lidar data,

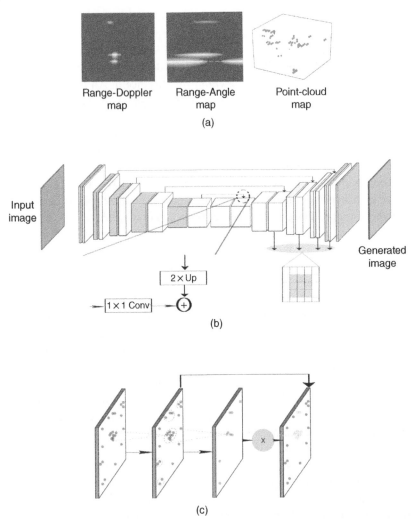

Figure 1.5 A visual summary on state-of-the-art target detection algorithms using (a) different target representation map followed by (b) convolution neural network architecture for image processing, and (c) PointNet architecture for processing of point-clouds.

radar data contains other attributes in the form of Doppler component or target radar cross section (RCS) information. As a result, multiple frameworks for target detection using radar point clouds were proposed [12–19]. The richness of data points for sparse radar point-clouds is improved either by accumulating the data over multiple frames or by augmenting data by repetition of target samples from available measurement by random sampling.

1.4 Target Tracking

Conventional radar signal processing usually applies a tracking algorithm after the clustering to filter measurements over time and track individual targets. False detections arising from ghost targets are usually also eliminated during the tracking. In order to avoid decreasing the measurement accuracy as well as the introduction of inherent noise, the usage of recursive filters is preferred in literature [20–27]. The most widely used tracking algorithms are Kalman filters. The performance of Kalman filters relies on the state vector (i.e., the parameters to be tracked), measurement noise, process noise, and the transition from measurement to the input state space. In the case of target detection, the tracker's state vector can be described as $\mathbf{x} = \begin{bmatrix} px & py & v & \theta \end{bmatrix}^T$, where $px, py, v,$ and θ are the position coordinates along the x- and y-axes, the radial velocity and the azimuth angle, respectively. Due to low variations in the spatial dimension, azimuth information is also used as part of the state vector and improves the robustness of the target localization. Generally, heading angle and turn-rate bring additional information for a dynamic target with nonlinear motion.

Once the target detections are available at each frame, the next step is to associate them with existing tracks or create or delete tracks as necessary. The detected targets and their estimated parameters are prone to errors due to measurement noise, missed detections due to occlusions, and false alarms due to ghost targets and interference sources. The task of tracking is to maintain and update the state and identity of these targets over time reducing such false positive and true negatives. There are two important aspects to tracking, first is the track management part, which handles track initiation, track maintenance, and measurement to track association. The other part is the track-filtering step, which is achieved through either alpha-beta filter, Kalman filter, or particle filter.

1.4.1 Track Management

Track management involves logic blocks that record target tracks when it needs to be initialized or confirmed, maintains or removes a track signifying exit of a target from the field-of-view and also handles the measurement to track association, which decides on the assignment of a detected target or measurement to an existing track or generation of a new track if sufficient criteria are fulfilled. The track management block acts as a controller to handle initiation, maintenance, and deletion of a target track.

Track initiation is the process of creating a new track from an unassociated radar measurement. During track initialization, when a new unassociated radar measurement is encountered, the measurement is assigned to a new track, but the status of the track is kept tentative. The status of these tracks is changed to confirmed only when N_{det} out of N_{fr} ($N_{det} \leq N_{fr}$) subsequent measurements have a positive detection/measurement of the target. This helps in reducing false positives from ghost targets or interference to spin off a target track.

Track maintenance is the process in which a decision is made if a track has to be terminated. If a track was not associated with a measurement during the association phase for consecutive N_{fr} frames, there is a high likelihood that the target no longer exists in the radar's field of view and thus such tracks are terminated. Alternately, due to missed detection, there are chances that the radar did not detect that target for few measurements within the N_{fr} frame window but would appear at future frame time. In this case, the tracks are retained, and only the prediction is updated.

The task of the measurement to track association is to assign the current measurement to an existing track. This step is also referred to as gating in literature. At time step $n + 1$, there are N tracks, i.e.,

$$X = \{(\mu_1, \phi_1), (\mu_2, \phi_2), \dots, (\mu_N, \phi_N)\} \tag{1.21}$$

where $\{\mu_i, \phi_i\}_{i=1}^N$ are the state mean and state error covariance matrices of the tracks. Additionally, the M measurements coming from target detections at time step $n + 1$ are

$$Z = \{z_1, z_2, \dots, z_M\} \tag{1.22}$$

A measurement of track association function $\Phi: Z \rightarrow X$ maps the uncertain measurements to certain tracks. The measurement to track association is a difficult problem due to multiple targets and the associated false positives and true negatives of the target detections. One of the simplest measurements to track association algorithm is suboptimal nearest neighbor (SNN), whereby each measurement is assigned to a track closest to it in Euclidean or Mahalanobis distance sense, and the assignment starts with the strongest detected target and thus is referred to as SNN. The alternate approach includes probabilistic data association filter or joint probabilistic data association filter.

1.4.2 Track Filtering

The Kalman filter is the most well-known track-filtering algorithm and utilizes a series of noisy measurements observed over time to estimate the internal state of a linear dynamic system. The estimated state variables are more accurate since the algorithm estimates a joint probability distribution over the variables for each time frame. The algorithm has two steps – prediction step and the measurement step. In the prediction step, the filter predicts the next internal state based on the process model accounting for the model uncertainties. While in the measurement step, the measurement is combined with the predicted state in an adaptive weighted average referred to as "Kalman gain" to update the internal state estimate and its uncertainty. The Kalman filter operates under Bayesian principle, so when the measurement data are observed with less uncertainty compared to the predicted data, the Kalman gain is high, which means it relies more on the measured data to update the state. On the other hand, when the predicted data have less uncertainty compared to that of the measurement data, the Kalman gain is low, meaning it weighs the predicted data more for its state update. Kalman filters are a generic algorithm which finds use in several signal processing to control theoretic applications and is not limited to only radar signal processing. The process model used by the Kalman filter is similar to that of a hidden Markov model except that the state space is rather continuous and assumes the state and observed variables to follow a normal distribution.

One of the most standard used process model or state transition function is a linear constant velocity (CV) motion model. Let the state variable be $\psi = \begin{bmatrix} p^x & p^y & v^x & v^y \end{bmatrix}^T$, where p^x and p^y denote the position coordinates in x and y coordinates, whereas v^x and v^y denote the velocity in the x- and y-axes, respectively. The CV process model can be expressed as follows:

$$
\begin{cases}
p^x(k) = p^x(k-1) + v^x(k-1)\delta t \\
p^y(k) = p^y(k-1) + v^y(k-1)\delta t \\
v^x(k) = v^x(k-1) + n^p_{vx} \\
v^y(k) = v^y(k-1) + n^p_{vy}
\end{cases}
\tag{1.23}
$$

where k is the time step, δt represents the frame time, i.e., the radar refresh rate. n^p denotes the Gaussian process noise and captures the kinematic changes which are not considered in the linear process update model. In the compact matrix–vector form, the process model can be expressed as follows:

$$
\begin{bmatrix}
p^x(k) \\
p^y(k) \\
v^x(k) \\
v^y(k)
\end{bmatrix}
=
\begin{bmatrix}
1 & 0 & \delta t & 0 \\
0 & 1 & 0 & \delta t \\
0 & 0 & 1 & 0 \\
0 & 0 & 0 & 1
\end{bmatrix}
\begin{bmatrix}
p^x(k-1) \\
p^y(k-1) \\
v^x(k-1) \\
v^y(k-1)
\end{bmatrix}
+
\begin{bmatrix}
0 \\
0 \\
n^p_{vx} \\
n^p_{vy}
\end{bmatrix}
\tag{1.24}
$$

or $\psi(k+1) = F\psi(k) + n^p$. Now, since the state variables are assumed to be Gaussian distributed, the process noise is modeled as zero mean Gaussian noise, and thus the mean and covariance of the state variables can be expressed as follows:

$$
\begin{aligned}
x_{k|k-1} &= \mathbf{E}\{\psi_{k|k-1}\} = Fx_{k|k-1} \\
P_{k|k-1} &= \mathbf{E}\{\phi_{k|k-1}\phi^T_{k|k-1}\} = FP_{k-1|k-1}F^T + Q
\end{aligned}
\tag{1.25}
$$

where the state variables with suffix $k-1|k-1$ denotes the posterior probability and $k|k-1$ denotes the prior probability, and $\mathbf{E}\{.\}$ is the expectation operator and Q denotes the process noise covariance matrix. In the case of CV model, the process noise would include kinematic changes due to jerks and acceleration of the target. Considering the acceleration component, the noise process can be expressed as follows:

$$
n^{\mathrm{p}} = \begin{bmatrix} \frac{1}{2}a_x\delta t^2 \\ \frac{1}{2}a_y\delta t^2 \\ a_x\delta t \\ a_y\delta t \end{bmatrix} = \begin{bmatrix} \frac{1}{2}\delta t^2 & 0 \\ 0 & \frac{1}{2}\delta t^2 \\ \delta t & 0 \\ 0 & \delta t \end{bmatrix} \begin{bmatrix} a_x \\ a_y \end{bmatrix} = Ga \tag{1.26}
$$

Thus, the process noise covariance matrix is defined as follows:

$$
Q = \mathbf{E}\{n^{\mathrm{p}}n^{\mathrm{p}T}\} = G \begin{bmatrix} a_x^2 & 0 \\ 0 & a_y^2 \end{bmatrix} G^T \tag{1.27}
$$

where a_x^2 and a_y^2 represent the variance of acceleration noise in x- and y-axes, and the covariance acceleration noise in x- and y-axes are 0. Thus, the process noise covariance Q is initialized based on the maximum target's acceleration expected in the system.

At the next step, the measurement model needs to account for the transformation of the internal state to the radar measurements, i.e., the radial distance ρ, the azimuth angle or bearing angle θ, and the radial velocity or range rate v.

$$
\begin{cases} \rho = \sqrt{p^{x2} + p^{y2}} \\ v = \frac{p^x v^x + p^y v^y}{\sqrt{p^{x2} + p^{y2}}} \\ \theta = \tan^{-1}\left(\frac{p^y}{p^x}\right) \end{cases} \tag{1.28}
$$

That is, $z = H(\psi) + n^m$, where n^m represents the measurement zero-mean Gaussian noise with covariance matrix $R = \mathbf{E}\{n^m n^{mT}\}$. This indicates that unlike the process model, the measurement model which maps the predicted state $\begin{bmatrix} p^x & p^y & v^x & v^y \end{bmatrix}^T$ to the measurement space $z = \begin{bmatrix} \rho & v & \theta \end{bmatrix}^T$ is a nonlinear transformation, which implies that the simple Kalman filter cannot be applied in this case. The two common approaches to handle such a nonlinear transformation issue are namely extended Kalman filter (EKF) and the unscented Kalman filter (UKF). While in EKF, the nonlinear equation is approximated through first-order Taylor's expansion. It approximates and propagates the state distribution through the first-order Taylor series linearization, which expands the nonlinear state around a single-point. As a result, the EKF is not able to capture the uncertainty of the distribution, introducing large errors in the estimation of the true posterior mean and covariance, respectively. Alternately, UKFs are used, which uses deterministic sampling filters, i.e., a sigma-point Kalman filter (SPKF), to approximate

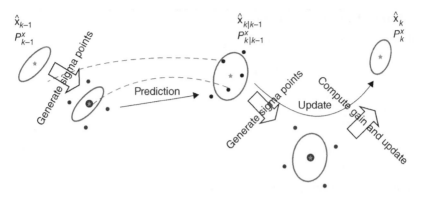

Figure 1.6 Graphical representation of predict and update stage for an UKF where mean and variance are estimated at each stage by approximation over sigma points.

the measurement model by a minimal set of sample points. These sample points can capture the true mean and covariance of the Gaussian random variable (GRV). Figure 1.6 gives a visual overview on the prediction and update operation of the UKF to track mean ($\hat{x}_{k-1}$) and covariance (P^x_{k-1}) of the input state vector at a time instance $k-1$, while Algorithm 1.1 gives a mathematical understanding on the implementation of the UKF.

In the UKF, an unscented transformation (UT) is applied at both prediction and update steps, which include a nonlinear state transformation based on f and h, respectively. As input, the state vector x_{k-1} of dimension n_x with mean $\hat{x}^{(i)}_{k|k-1}$ and the covariance $P^x_{k|k-1}$ are provided. At prediction stage, sigma points $\hat{x}^{(i)}_{k-1|k-1}$ are generated, which undergoes the UT $f(.)$ to estimate the predicted mean $\hat{x}_{k|k-1}$ and the covariance $P^x_{k|k-1}$ of the state vector. Since the predicted mean and variance changed due to the transformation, a new set of sigma points is calculated. Afterward, the new sigma points are transformed into measurement space using $h(.)$ as a transformation function. A CV system is considered with the localization state vector **x**. A nonlinear measurement model $h(.)$ accounts for the transformation of the state vector into the measurement domain. Mapping part of the localization parameters (radial range and azimuth angle) from the tracker's state vector to the measurement domain follows a nonlinearity (Cartesian to spherical), whereas mapping the radial velocity and augmented parameters (appearance embedding) corresponds to an identity mapping between state vector and measurement domain. However, the overall nonlinear transformation in the process model $x^P = g(x_a)$ and the measurement model $z^P = h(x^P_a)$ can be achieved through the UT, using the so-called "sigma points." These are generated to approximate the statistical properties of the state distribution [28].

Algorithm 1.1 Unscented Kalman filter.

Prediction: Generate sigma points $\hat{x}_{k-1|k-1}^{(i)}, i = 0, 1, \ldots, 2n_x$

$$\hat{x}_{k|k-1}^{(i)} = f\left(\hat{x}_{k-1|k-1}^{(i)}\right), i = 0, 1, \ldots, 2n_x,$$

$$\hat{x}_{k|k-1} = \sum_{i=0}^{2n} W_i^{(m)} \hat{x}_{k|k-1}^{(i)},$$

$$P_{k|k-1}^x = \sum_{i=0}^{2n_x} W_i^{(c)} \left(\hat{x}_{k|k-1}^{(i)} - \hat{x}_{k|k-1}\right)\left(\hat{x}_{k|k-1}^{(i)} - \hat{x}_{k|k-1}\right)^T + Q$$

Update: Generate sigma points $\hat{x}_{k|k-1}^{(i)}, i = 0, 1, \ldots, 2n_x$

$$\hat{y}_{k|k-1}^{(i)} = h\left(\hat{x}_{k|k-1}^{(i)}\right), i = 0, 1, \ldots, 2n_x,$$

$$\hat{y}_{k|k-1} = \sum_{i=0}^{2n_x} W_i^{(m)} \hat{y}_{k|k-1}^{(i)},$$

$$P_{k|k-1}^y = \sum_{i=0}^{2n_x} W_i^{(c)} \left(\hat{y}_{k|k-1}^{(i)} - \hat{y}_{k|k-1}\right)\left(\hat{y}_{k|k-1}^{(i)} - \hat{y}_{k|k-1}\right)^T + R,$$

$$P_{k|k-1}^{xy} = \sum_{i=0}^{2n_x} W_i^{(c)} \left(\hat{x}_{k|k-1}^{(i)} - \hat{x}_{k|k-1}\right)\left(\hat{y}_{k|k-1}^{(i)} - \hat{y}_{k|k-1}\right)^T,$$

$$\hat{x}_{k|k} = \hat{x}_{k|k-1} + P_{k|k-1}^{xy}\left(P_{k|k-1}^y\right)^{-1}\left(y_k - \hat{y}_{k|k-1}\right),$$

$$P_{k|k}^x = P_{k|k-1}^x - P_{k|k-1}^{xy}\left(P_{k|k-1}^y\right)^{-1}\left(P_{k|k-1}^{xy}\right)^T.$$

1.5 Target Representation

Unlike computer vision, target recognition for radar sensors rely on the target representation, which further depends on the choice of frequency transforms. As discussed before in Sections 1.1.2 and 1.2.1, both signal modulation and frequency-transform parameters are critical to specify the required resolution for any target representation for an application. While the signal modulation parameter defines both maximum and minimum measurable range-Doppler quantities, the frequency transform parameters are used to find a trade-off between time-frequency resolution of target's spatial and motion signatures. In broad terms, target recognition is achieved by target detection, tracking, and feature extraction for classification. As a result, accuracy of target classification strongly depends on accurate detection and tracking of the target.

Unlike camera-based imaging, radar images appear dense in nature, but target information is very low compared to background or noise distribution. Similarly, in comparison to LiDAR point-clouds, where density of target point-clouds are richer, radar point-clouds are very sparse for a single frame which makes it hard for target detection and thus more complex for target classification. In case of radar-point clouds, as a most common practice, multiple frames are accumulated over time to make target representation richer. The overview on state-of-the-art target representations in the context of target recognition is summarized in the following sections.

1.5.1 Image Representation

Radar feature images are 2D representations of the target that can be used to extract additional information about the target. Based on the radar images, the type of the target, e.g., if the target is a pedestrian, a car, or an animal, or the currently executed activity such as standing idle, walking, or riding a bike, can be recognized. The relevant 2D radar features that can be extracted for target classification are Doppler spectrograms, range angle images, and video of RDIs, which are presented in the sequel.

1.5.1.1 Doppler Spectrogram

The Doppler spectrogram is generated by first detecting the target along the range bins, followed by FFT along slow-time for the detected range bin. The Doppler spectra from consecutive frames are stacked one after another to generate a 2D image. In the case of a single target, the Doppler spectrum can be achieved by marginalization of the RDI across the range axis. The Doppler spectrum contains both the macro-Doppler component as well as micro-Doppler components due to hand and leg movements while performing an activity. The stacked Doppler spectrum across consecutive frames is referred to as Doppler spectrogram that captures information about the instantaneous Doppler spectral content and the variation of the Doppler spectral content over time. The Doppler spectra of the slow-time data from kth radar frame on the selected range bins can be expressed as follows:

$$S(p,k)^n = \left| \sum_{m=1}^{N_{st}} w(m)s(m,k)^n \exp\left(-\frac{j2\pi mp}{N_{st}}\right) \right|$$

$$S(p,k) = \sum_{n=1}^{N} S(p,k)^n \tag{1.29}$$

where $w(m)$ is the window function along slow-time indexed by m, $s(m,k)^n$ is the slow-time data across N_c chirps in the kth frame for the nth range bin, and p is the FFT points along slow-time. $S(p,k)^n$ represents the Doppler spectrogram along the nth range bin, and N represents the range bins along which the Doppler FFT

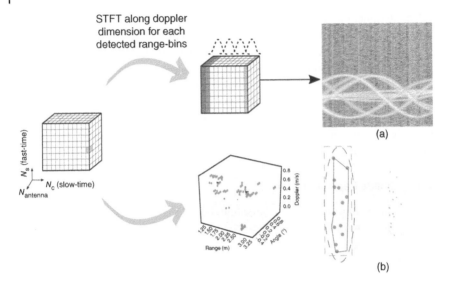

Figure 1.7 Target classification using (a) Doppler signature image and (b) spatial point-cloud maps.

is summed to generate the Doppler spectra for that frame. Figure 1.7 illustrates an example of typical motion signature of a cyclist which is generated using STFT, as described in Section 1.2.1.

1.5.1.2 Range Angle Images

Two receive antennas separated by a distance of d and receiving reflections from angle θ implies the second antenna receives the reflected signal with an additional path length of $d\sin(\theta)$ as shown in Figure 1.8 leading to phase difference of

$$\delta\phi = \left(\frac{2\pi}{\lambda}\right) d\sin(\theta) \tag{1.30}$$

Thus, the angle of arrival θ can be estimated as follows:

$$\hat{\theta} = \sin^{-1}\left(\frac{\lambda\delta\phi}{2\pi d}\right) \tag{1.31}$$

This angle of arrival estimation is referred to as phase monopulse technique.

Using the same concept, digital beamforming generates the weights for predefined angles sweeping across the entire angular field of view to create range-angle image, where the peak in the spectrum denotes the angle of arrival of the target at that particular range. Similarly, there are several other approaches to obtain the range angle image. A simplest approach takes an FFT along the virtual channel after applying a window function and zero-padding.

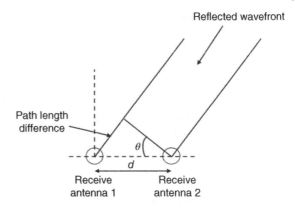

Figure 1.8 Two receive antennas with the angle of arrival θ and the path length difference between the two.

The other approach is referred to as Capon beamforming or minimum variance unbiased estimator. The $N_t \times N_r$ de-ramped beat signal can be stacked into a vector and the Kronecker product of the steering vector of the Tx array $a^{Tx}(\theta)$ and the steering vector of the Rx array $a^{Rx}(\theta)$, i.e., $a^{Tx}(\theta) \otimes a^{Rx}(\theta)$ can be used to resolve the relative angle θ of the scatterer, where $\otimes$ represents the Kronecker product. Subsequently beamforming of the MIMO array signals can be regarded as synthesizing the received signals with the Tx and Rx steering vectors. The azimuth imaging profile for a range bin l can be generated using the Capon spectrum from the beamformer. The Capon beamformer is computed by minimizing the variance/power of noise while maintaining a distortion-less response toward a desired angle. The corresponding quadratic optimization problem is

$$\min_{w} w^H C w$$
$$s.t. \quad w^H (a^{Tx}(\theta) \otimes a^{Rx}(\theta)) = 1 \tag{1.32}$$

where C is the covariance matrix of noise, and the above optimization has a closed form expression given as $w_{capon} = \frac{C^{-1} a(\theta)}{a^H(\theta) C^{-1} a(\theta)}$, with θ being a desired angle. On substituting w_{capon} in the objective function, the spatial spectrum is given as follows:

$$P_l(\theta) = \frac{1}{\left(a^{Tx}(\theta) \otimes a^{Rx}(\theta)\right)^H C_l^{-1} \left(a^{Tx}(\theta) \otimes a^{Rx}(\theta)\right)} \tag{1.33}$$
$$\text{with } l = 0, ..., L$$

However, estimation of noise covariance at each range bin l is difficult in practice; hence, $\hat{C}_l$ is estimated which contains the signal component as well and can be estimated using sample matrix inversion technique $\hat{C}_l = \frac{1}{N} \sum_{k=1}^{K} s_l^{IF}(k) s_l^{IF}(k)^H$, where K denotes the number of snapshot used for signal plus noise covariance estimation and $s_l^{IF}(k)$ is the de-ramped IF signal at range bin l with k being the frame index.

1.5.1.3 Video of Range-Doppler Images

The video of RDIs computed at frame time k can be expressed as follows:

$$v_{RDI}(p, l, k) = \left| \sum_{m=1}^{N_{st}} \sum_{n=1}^{N_{ft}} w(m, n)s(m, n, k) \exp\left(-j2\pi\left(\frac{mp}{N_{st}} + \frac{nl}{N_{ft}}\right)\right)\right| \quad (1.34)$$

where $w(m, n)$ is the 2D weighting function along the fast-time and slow-time, while $s(m, n, k)$ is the ADC data on kth frame. The indices n, m sweep along the fast-time and slow-time axis, respectively, while l, p sweep along the range and Doppler axes, respectively. N_{st} and N_{ft} are the FFT size along the slow-time and fast-time, respectively.

1.5.2 Point-Cloud Maps

In comparison to images, the radar point-cloud representation enables a target classification framework, as shown in Figure 1.7b. The processing of point-cloud target classification follows a principle similar to image-based target classification where first neighborhood selection of all points is done using a clustering technique. The most common techniques used are DBSCAN. Thereafter, hand-crafted feature extraction and selection are performed, which is passed to linear or nonlinear supervised classifier. The hand-crafted feature engineering is mainly based on point-cloud distribution within the target cluster along each dimension. This includes mean value, maximum–minimum value within the cluster, and eigenvalues of the covariance matrix of x–y coordinates of clustered target points. Additionally, concepts like mean-shift estimation for each point or entire cluster also help to make point cloud features more robust. Additionally, with availability of deep learning frameworks [9–11] to process point-clouds directly, enable an end-to-end framework for feature engineering and classification. The most common practices are as follows:

- Processing point-cloud maps directly by PointNet to generate a class score for each point which can be clustered into one class.
- Processing point-cloud clusters separately instead of maps directly.
- Mapping point-clouds to a global coordinate grid map and using vision-based detector network architectures like you-only look once (YOLO) [29], single shot detector (SSD) [30], and region-based fully convolutional network (RFCN) [31].
- Combining PointNet features with a recurrent neural network (RNN) for temporal smoothening.

Similar to images, target classification using radar point-clouds is very challenging due to inherent nature of data distribution. Although the point-cloud maps contain information such as angle, range, velocity, RCS, orientation, and dimension for each detected target point, the point distribution is very sparse in nature. Further, the number of data points for weak targets such as pedestrian

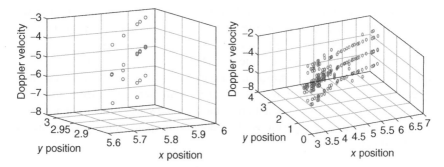

Figure 1.9 Illustration of a 3D radar point cloud where coordinate axis are *x*, *y* positional coordinates and Doppler velocity; (a) and (b) show how data aggregation changes the form and information content of a point cloud.

and cyclist are in the range of 0–10 within a single frame, depending on the target's viewing angle and distance to radar sensor. Further, this also creates a data imbalance within the data distribution of targets despite of having the same number of examples for each target of interest. Additionally, radar point-cloud maps are very prone to viewpoint and temporal variations across consecutive measurements of the same object even during a static scenario. As an alternative to this problem, multiple frames are accumulated. These variations occur due to the radar sensors statistical nature and due to the use of the classical radar processing, which uses handcrafted features (e.g., detector thresholds) to detect and localize radar detections. As an alternative to it, multiple frames are accumulated to generate dense point-cloud maps. Figure 1.9 illustrates how the point cloud changes due to aggregation of multiple point-clouds over time.

1.6 Target Recognition

After accurate target localization using detection and tracking along either of two dimensions, i.e., range, Doppler, and angle, detected target is classified into desired class. Similar to detection, multiple frameworks are proposed for target classification using both images and point-clouds. The most common approach for target classification using images is done by extracting hand-engineered features from detected target characteristic along range-angle and range-Doppler dimensions [32–34]. The most common characteristics are mean value of range, Doppler, direction of arrival, and normalized reflected power for all detected peaks from unique target cluster. Additionally, addition of variance and deviation in range, radial Doppler, and object size in *x,y*-dimension in feature set increase the richness of target features. Alternative to hand-crafted features, feature descriptors such as speed up robust features (SURFs) or scale-invariant feature transform (SIFT) can be used over the detected target region for a fixed grid size. Later, these feature set can be passed to a linear or nonlinear classifiers such as decision tree, support vector machine (SVM), or K-nearest neighbor (KNN). Such methods are very specific to sensor measurement and are prone to noise, and thus often, struggle to generalize over different measurements for sensor at diverse location or with different gain pattern. The alternate state-of-the-art approaches involve deep learning algorithms. The notable deep learning architecture is presented in the sequel.

1.6.1 Feedforward Network

Feedforward neural networks or multilayer perceptrons (MLPs) approximate some function f that does the mapping $y = f(x; w)$ and learns the weights w of the network that results in the function approximation given a certain objective. These models are called feedforward because information flows through the function being evaluated from x, through the intermediate computations used to define f, and finally to the output y. There are no feedback connections in which outputs of the model are fed back into itself. Figure 1.10 presents a feedforward network with one hidden layer.

1.6.2 Convolutional Neural Networks (CNN)

In deep learning literature, the most widely used networks are CNNs [35–37] for tasks such as object detection, face recognition, image segmentation, or super-resolution. In CNNs, the image classification is performed by incorporating various layers namely convolution layers, pooling layers, and dense layers with cross-entropy (CE) loss. The network sees an image as a multidimensional array of pixels with the dimensions $h \times w \times c$ (h = height, w = width, c = color channel).

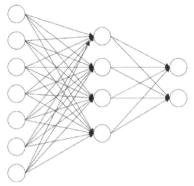

Figure 1.10 Example of a feedforward neural network with one hidden layer.

Input layer $\in R^7$ Hidden layer $\in R^4$ Output layer $\in R^2$

As an example, an image of size $32 \times 32 \times 3$ represents an RGB image (three-color channels), while an image of size $32 \times 32 \times 1$ represents a grayscale image.

The initial spatial feature extraction in a CNN is done by convolution layers. In a convolution layer, a filter or a kernel (weight matrix) is sliding through the entire image. The output of the convolution layer is defined as the dot product of the sliding kernel and the underlying input image in each sliding step. One must note that the filter must have the same number of channels as in the input image.

The output dimension of the convolution layers are calculated as follows:

$$w_{out} = \frac{w_{in} - f_w + 2p}{s} + 1$$

$$h_{out} = \frac{h_{in} - f_h + 2p}{s} + 1 \tag{1.35}$$

where w_{in}, h_{in} are the width and the height of the input image, f_w and f_h are the width and height of the filter or kernel, p, s are the padding and the stride factors and are set ≥ 1. The weights of the filters are learned by the network through back propagation. The components of the convolution layers are described as follows:

1. Strides: While applying convolution, shifts are made to move the filter across the entire image. The stride defines the step size of the shift, e.g., if the stride is one, then the filter is shifted by a single pixel, and if the stride is two, then it is shifted by two pixels, and so on.
2. Padding: When a filter does not fit the input image properly, there are two options:
   - Zero-padding – Pad the image with zeros so that the filter fits perfectly.
   - Valid-padding – Dropping the part of images that did not fit the filter perfectly.
3. Activation Function: Some of the standard activation functions used in a CNN are as follows:
   - Sigmoid (Logistic Activation): This activation function is originally inspired from the "real neuron." The output of this activation function is between [0,1]. Its major drawbacks are that a saturated neuron will not learn and the activation is computationally expensive.

$$f(x) = \frac{1}{1 + \exp(-x)}$$

$$f'(x) = f(x)(1 - f(x)) \tag{1.36}$$

   - Hyperbolic Tangent Activation: The output range of hyperbolic tangent activation is between $[-1,1]$ and is zero centered. Similarly like sigmoid activation, this activation also does not train saturated neurons.

$$f(x) = \tan h(x) = \frac{1 - \exp(-x)}{1 + \exp(-x)}$$

$$f'(x) = 1 - \tan h^2(x) \tag{1.37}$$

- Rectified Linear Unit (ReLu): In ReLu activation, there is no saturation if $x > 0$ and is more computationally efficient and leads to faster convergence. In such an activation, the output is always positive and the inactive neurons are not optimized:

$$f(x) = \max(0, x)$$

$$f'(x) = \begin{cases} 1 & \text{if } x > 0 \\ 0 & \text{if } x < 0 \end{cases} \tag{1.38}$$

- Leaky Rectified Linear Unit and Parametric ReLu: These activation functions are an improvement over the normal ReLu which overcomes the problem of dead neurons. In case of leaky Relu, α is a small constant, such as 0.01. In case of parametric ReLu, α is a hyperparameter learned through back propagation.

$$f(x) = \begin{cases} x & \text{if } x > 0 \\ \alpha x & \text{if } x < 0 \end{cases}$$

$$f'(x) = \begin{cases} 1 & \text{if } x > 0 \\ \alpha & \text{if } x < 0 \end{cases} \tag{1.39}$$

Figure 1.11 presents the commonly used activation functions described above namely sigmoid function, hyperbolic function, rectified linear unit (ReLu), and leaky rectified linear unit.

One of the other standard activation function typically used in the output layer for classification problems is the softmax layer. Since the squared error is not suitable for cases where classes are mutually exclusive, a better approach is to assign probabilities to each class with the constraint that the outputs should sum up to 1. The softmax function forces the output to represent a probability distribution across the possible classes L. Its function and its derivatives are given as follows:

$$p_k = \frac{\exp(x_k)}{\sum_{j=1}^{L} \exp(x_j)}$$

$$p'_k = f(x_k)(1 - f(x_k)) \tag{1.40}$$

The cost function typically associated with the softmax layer is the negative log likelihood of the correct prediction called CE or log loss cost function and is defined as follows:

$$E(t, p) = -\sum_{k=1}^{L} t_k \log\left(\frac{\exp(u_k)}{\sum_{j=1}^{L} \exp(u_j)}\right) \tag{1.41}$$

4. Pooling/Subsample Layers: Pooling layers or more specifically spatial pooling layers perform subsampling or down sampling of the input image while retaining the most relevant information. This process helps in the reduction of parameters when the images are too large.

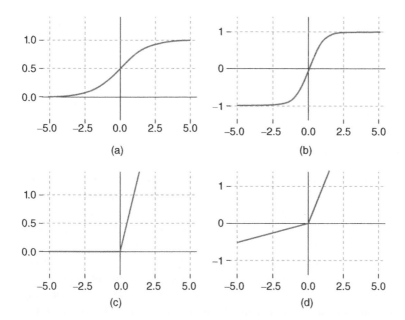

Figure 1.11 Illustration of various activation functions: (a) sigmoid function, (b) hyperbolic function, (c) ReLu function, and (d) leaky ReLu.

Three majorly used pooling layers are as follows:
- Max pooling – Taking the largest element within the defined nonparametric filter size.
- Average pooling – Taking the average of all the elements within the defined nonparametric filter size.
- Sum pooling – Taking the sum of all the elements within the defined nonparametric filter size.
5. Dense/Fully Connected Layers: At the end of the CNN, a single or multiple dense layers are used to which the flattened (1D array) output of the previous convolution and pooling layers is fed. In a CNN used for classification the last activation function is typically a sigmoid or a softmax activation.

Figure 1.12 illustrates an example of a CNN architecture that includes convolution layers, pooling layers, or subsample layers followed by dense or fully connected layers at the later stage.

1.6.3 Recurrent Neural Network (RNN)

MLPs and CNNs cannot directly address the problem of information propagation through time. Several applications such as gesture sensing and tracking require

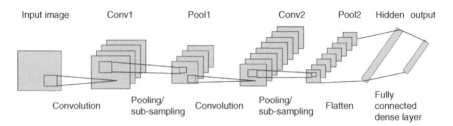

Input image   Conv1   Pool1   Conv2   Pool2   Hidden output

Convolution   Pooling/sub-sampling   Convolution   Pooling/sub-sampling   Flatten   Fully connected dense layer

Figure 1.12 Example of a CNN architecture.

the neural network to keep a history of the past events to make a decision. RNN addresses this issue, by having self-loop structures, allowing information to persist over time. Using these self-loops, RNNs are able to link previous information to the current task and make a decision based on the previous events. One of the biggest hindrance toward wide adoption of RNNs in 1990s was the problem of vanishing gradient. Not only does the information flows through time but also the error has to be backpropageted through time. To do so, the self-loops are unfolded over time, which leads to a very deep network, where the gradient has to propagate through many layers. However, if the weights are less than one, further multiplication by gradient that is also less than one would result in a very small number after a few multiplications. Thus, the gradient flows over time through the RNNs would easily become zero, which means no further propagation of information. As a result, the RNNs were unable to retain information or learn information from quite distant past.

RNNs can be described as follows: a RNN maps a given temporal input sequence $\mathbf{x}(k) = (x_1(k), x_2(k), x_3(k))$ to a sequence of hidden values $\mathbf{h}(k) = (h_1(k), \ldots, h_T(k))$ and outputs a sequence of activations $\mathbf{a}(k+1) = (a_1(k+1), \ldots, a_T(k+1))$ by iterating the following recursive equation:

$$\mathbf{h}(k) = \sigma(W_{hx}\mathbf{x}(k) + \mathbf{h}(k-1)W_{hh} + \mathbf{b}_h) \tag{1.42}$$

where σ is the nonlinear activation function, $\mathbf{b}_h$ is the hidden bias vector, W_{hx} is the input-hidden weight matrix and W_{hh} is the hidden-hidden weight matrix.

The activation for these recurrent units is defined as follows:

$$\mathbf{a}(k+1) = h(k)W_{ha} + \mathbf{b}_a \tag{1.43}$$

where W_{ha} denotes the hidden-activation weight matrix and $\mathbf{b}_a$ denotes the activation bias vector.

RNNs have the problem of vanishing or exploding gradient, which is solved through a long-short term memory (LSTM) [38] or a gated recurrent unit (GRU) [39]. LSTMs extend RNNs with memory cells using the concept of gating: a mechanism based on componentwise multiplication of the input, which defines the behavior of each individual memory cell. The LSTM updates its cell state, according to the activation of the gates. The input provided to an LSTM is fed into different gates that control which operation is performed on the cell memory: write (input gate), read (output gate), or reset (forget gate). These gates act on the signals they receive and block or pass information based on its strength and importance, by virtue of their own learned filter weights. These weights are learned during backpropagation, which means the weights of the cells decide when to allow data to be entered, to be retained, or to be deleted.

The vectorial representation (vectors denoting all units in a layer) of the update of an LSTM layer is as follows:

$$
\begin{cases}
\mathbf{i}(k) = \sigma_i(W_{ai}\mathbf{a}(k) + W_{hi}\mathbf{h}(k-1) + W_{ci}\mathbf{c}(k-1) + \mathbf{b}_i) \\
\mathbf{f}(k) = \sigma_f(W_{af}\mathbf{a}(k) + W_{hf}\mathbf{h}(k-1 + W_{cf}\mathbf{c}(k-1) + \mathbf{b}_f) \\
\mathbf{c}(k) = \mathbf{f}(k)\mathbf{c}(k-1) + \mathbf{i}(k)\sigma_c(W_{ac}\mathbf{a}(k) + W_{hc}\mathbf{h}(k-1) + \mathbf{b}_c) \\
\mathbf{o}(k) = \sigma_o(W_{ao}\mathbf{a}(k) + W_{ho}\mathbf{h}(k-1) + W_{co}\mathbf{c}(k) + \mathbf{b}_o) \\
\mathbf{h}(k) = \mathbf{o}(k)\sigma_h(\mathbf{c}(k))
\end{cases}
\tag{1.44}
$$

where $\mathbf{i}$, $\mathbf{f}$, $\mathbf{o}$, and $\mathbf{c}$ are the input gate, forget gate, output gate, and cell activation vectors, respectively, all of which are of the same size as vector $\mathbf{h}$ defining the hidden value. Terms σ represent nonlinear activation functions. The term $\{\mathbf{x}(1), \mathbf{x}(2), \cdots , \mathbf{x}(K)\}$ is the input to the memory cell layer at time k. W_{ai}, W_{hi}, W_{ci}, $W_{af}, W_{hf}, W_{cf}, W_{ac}, W_{hc}, W_{ao}, W_{ho}$, and W_{co} are weight matrices, with subscripts representing from-to relationships b_i, b_f, b_c, and b_o are bias vectors.

Figure 1.13 illustrates one unit of a LSTM block. Depending on the application, there are different configurations how the RNN models can be used. RNNs are not only able to map one input to one output but can also map one input to multiple outputs, multiple inputs to one output or multiple inputs to multiple outputs.

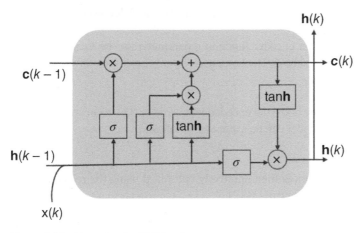

Figure 1.13 Example of a LSTM cell.

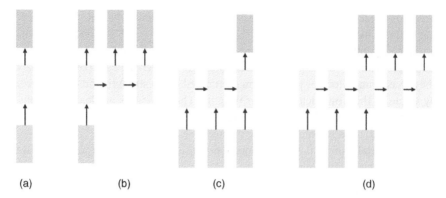

(a)   (b)   (c)   (d)

Figure 1.14 Different types of RNN models: (a) one-to-one, (b) one-to-many, (c) many-to-one, and (d) many-to-many.

These four configurations are illustrated in Figure 1.14. The many-to-one configuration is used in radar-gesture sensing and many-to-many configuration is used in human activity classification.

1.6.4 Autoencoder and Variational Autoencoder

Autoencoders comprise an encoder neural network followed by a decoder neural network with the aim of reconstructing the input data at the output. The design of the autoencoder imposes a bottleneck in the network that encourages a compressed representation of the original input. In general, autoencoders aim to leverage the key structure in the data to compress the input into the network's bottleneck or latent space representation, which is enough to reconstruct the original input data. It is thus used in dimension reduction and denoising applications among others.

The model involves an encoder function g parameterized by θ and a decoder function f parameterized by ϕ. The bottleneck layer is given as follows:

$$z = g(x; \theta) \tag{1.45}$$

where x is the input data and z the encoded latent vector. The reconstructed input at the output of the decoder can be expressed as follows:

$$\hat{x} = f(g(x; \theta); \phi) \tag{1.46}$$

The autoencoder network is then iteratively optimized using the reconstruction loss such as mean squared error (MSE):

$$L(\theta, \phi) = \frac{1}{N} \sum_{i=1}^{N} |x^i - \hat{x}^i|^2 \tag{1.47}$$

Compared to autoencoders that aim to project the input data onto a single latent vector, variational autoencoders [40–42] instead aim to learn or encode the input data onto a distribution in the latent space. Variational autoencoders can be viewed as applying a regularization during training that prevents the network from over-fitting. The input data x is encoded as a distribution over the latent space, i.e., $p_\theta(z|x)$, then a point is sampled from that latent distribution $z \sim p_\theta(z|x)$, which is then fed into the decoder to reconstruct the input data as $\hat{x} = g_\phi(z)$. The reconstruction loss such as the mean-square error is used along with the Kullback–Leibler (KL) divergence of $p_\theta(z|x)$ to a Gaussian distribution with mean 0 and variance 1, i.e., $\mathcal{N}(0,1)$, to back propagate and learn the weights of the network.

In practice, the encoded distributions are chosen to be normal so that the encoder can be trained to return the mean and the covariance matrix that describes these Gaussians. The reason why an input is encoded as a distribution with some variance instead of a single point is that it allows to express very naturally the latent space regularization: the distributions returned by the encoder are enforced to be close to a standard normal distribution such that the entire feature space is close to a standard normal distribution. We can notice that the KL divergence between two Gaussian distributions has a closed form that can be directly expressed in terms of the means and the covariance matrices of the two distributions. The loss function of a variational auto-encoder (VAE) can be written as follows:

$$L(\theta, \phi) = \frac{1}{N} \sum_{i=1}^{N} |x^i - \hat{x}^i|^2 + KL\left(p_\theta(z|x), \mathcal{N}(0,1)\right) \tag{1.48}$$

where N is the number of examples.

The KL divergence is the expectation of the log difference between the probability of data sampled from the approximating distribution and the target distribution and thus is defined as follows:

$$D_{\mathrm{KL}}(p\|q) = \sum_{i=1}^{N} p(x_i) \log\left(\frac{p(x_i)}{q(x_i)}\right) \tag{1.49}$$

The KL divergence has the following properties:

1. KL divergence is 0 when both distributions are approximately the same:

$$D_{\mathrm{KL}}(p\|q) = 0 \quad \text{iff} \quad p \sim q \tag{1.50}$$

2. KL divergence is always positive for any two distributions:

$$D_{\mathrm{KL}}(p\|q) > 0 \quad \text{if} \quad p \neq q \tag{1.51}$$

3. To ensure $D_{\mathrm{KL}}(p\|q)$ is finite, the support of p needs to be contained in q else if by Eq. (1.49) $q(x) \to 0$, then $D_{\mathrm{KL}}(p\|q) \to \infty$.

4. The KL divergence is an asymmetric metric, i.e.,

$$D_{KL}(p\|q) \neq D_{KL}(q\|p) \tag{1.52}$$

which means that $D_{KL}(p\|q)$ is not a distance metric

Conceptually, the VAE architecture of learning a distribution in the latent space makes the space continuous, meaning two closely spaced points in the latent space generate more similar content than two far spaced points, and complete, meaning any point sampled from the latent space generates a meaningful output at the decoder of the VAE. Since during backpropagation the gradient cannot flow through a probabilistic layer, the sampling process of extracting $z \sim p(z|x)$ requires a special technique, referred to as "reparameterization trick." The reparameterization trick suggests to randomly sample ϵ from a zero mean and unit variance Gaussian, and then shift ϵ by the latent distribution's mean μ and scale it by the latent distribution's variance σ. Figure 1.15 presents the reparameterization trick used to sample the random variable from the latent distribution making it deterministic. The reparameterization trick allows to optimize the parameters of the distribution while still maintaining the ability to randomly sample from that distribution.

1.6.5 Generative Adversarial Network

Generative adversarial networks (GANs) introduced by Goodfellow in 2014 [43] are a breakthrough in the field of unsupervised learning using neural

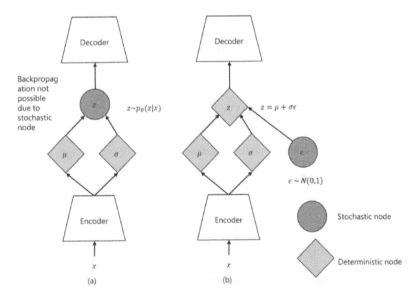

Figure 1.15 Illustration of variational autoencoder architecture depicting: (a) original form highlighting the issue during backpropagation, and (b) reparameterization trick.

networks. The technique is one of the most promising unsupervised learning approaches due to its capability of modeling high-dimensional distributions and less computationally expensive training process when compared with previous unsupervised learning methods such as VAEs, Boltzmann machines, and others. The working principle of GANs is a two-player minmax game, where two neural networks called the generator and discriminator are playing against each other. The generator tries to fool the discriminator by generating real-looking data, while the discriminator's task is to classify real and fake data. During training, the generator progressively becomes better at creating images that look real, while the discriminator becomes better at telling them apart. The minmax game has a global (and unique) optimum for $p_g = p_r$, where p_g is the generative distribution and p_r is the real data distribution. The process reaches equilibrium when the discriminator can no longer distinguish real from fake images. Once trained, only the generator is used to generate new realistic data similar to the real data distribution. Figure 1.16 illustrates the operating principle of a generator and discriminator that are used in training a vanilla GAN network.

During training, the discriminator classifies both real data and fake data from the generator and penalizes the discriminator weights for misclassifying a real instance as fake or a fake instance as real. Thus, incrementally getting better at classifying real and fake data. The generator part of a GAN learns to create fake data by incorporating feedback from the discriminator, in the sense that the generator loss penalizes the generator for failing to fool the discriminator. If the generator succeeds perfectly, then the discriminator has a 50% accuracy meaning it is unable to tell real from fake data anymore. If the GAN continues training past this point, then the generator starts to train on completely random feedback, and its own quality may collapse.

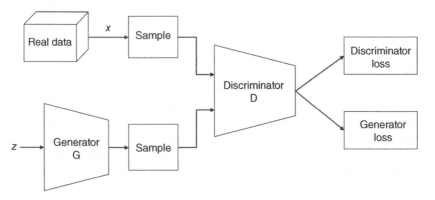

Figure 1.16 Illustration of a vanilla GAN architecture outlining the principle of a generator and a discriminator.

1.6.5.1 Minmax Loss

In the case of minmax loss, the objective of the discriminator is to maximize the expectation of the log likelihood of data drawn from the real distribution, i.e., $\max_D \mathbb{E}_{x\sim\mathbb{P}_r} \log(D(x))$, while minimizing the expectation of the log likelihood of data generated from the generator which samples from the random distribution $\mathbb{P}_z$, i.e., $\min_D \mathbb{E}_{z\sim\mathbb{P}_z} \log(D(G(z)))$ or equivalently $\max_D \mathbb{E}_{z\sim\mathbb{P}_z} \log(1 - D(G(z)))$. Thus, the objective of the discriminator function is

$$\max_D \mathbb{E}_{x\sim\mathbb{P}_r} \log(D(x)) + \max_D \mathbb{E}_{z\sim\mathbb{P}_z} \log(1 - D(G(z))) \tag{1.53}$$

On the other hand, the objective of the generator is to $\min_G \mathbb{E}_{z\sim\mathbb{P}_z} \log(1 - D(G(z)))$ such that the fake examples produced by the generator resemble the real data at the output of the discriminator. Thus, combining the two aspects and competing objective can be formulated as D and G are playing minmax game with the combined loss function:

$$\min_G \max_D [\mathbb{E}_{x\sim\mathbb{P}_r} \log(D(x)) + \mathbb{E}_{z\sim\mathbb{P}_z} \log(1 - D(G(z)))] \tag{1.54}$$

This is fine since $\mathbb{E}_{x\sim\mathbb{P}_r} \log(D(x))$ is independent of the generator optimization. It can be shown that the generator is trying to minimize the Jensen–Shannon (JS) divergence between $\mathbb{P}_r$ and $\mathbb{P}_g$. The JS divergence is bounded between 0 and 1 and is defined as follows:

$$D_{JS}(p\|q) = \frac{1}{2}D_{KL}\left(p\|\frac{p+q}{2}\right) + \frac{1}{2}D_{KL}\left(q\|\frac{p+q}{2}\right) \tag{1.55}$$

It is worth noting that unlike the KL divergence used in VAEs, the JS divergence is symmetric and in case of two distributions being disjoint would result in a maximum value of $\log(2)$, irrespective of the two distributions. In comparison, the KL divergence would be ∞ in such a scenario. From Eq. (1.55), it is easy to see that the minimum value of $D_{JS}(p\|q)$ is obtained when $p \sim q$. Consequently, the generator is trying to achieve $\mathbb{P}_g \sim \mathbb{P}_r$, which means that the generator generates data that resemble the real data. The discriminator maximizes the loss by trying to approach $D(x)$ to 1 and $D(G(z))$ to 0, thus attaining the optimal value of $D^*(x) = \frac{1}{2}$, which is the Nash equilibrium.

The minmax loss for GAN suffers from vanishing gradients and mode collapse. If the discriminator is too good, then the generator training can fail due to vanishing gradients. Furthermore, the generator in a random input GAN is expected to generate a variety of outputs. However, if a generator produces an especially plausible output, the generator may learn to produce only that output. If the generator starts producing the same output over several iterations, then the discriminator's best strategy is to reject that output always. But if the next iteration of discriminator gets stuck in a local minimum and does not find the best strategy, then it is too easy for the next generator iteration to find the most plausible output for the current discriminator. As a result, the generator gets trapped in a local minimum and generates limited set of outputs, and this phenomenon is referred as mode collapse.

1.6.5.2 Wasserstein Loss

In the Wasserstein generative adversarial networks (WGANs), the discriminator does not classify input data as real or fake, but rather outputs a number. Discriminator training just tries to make the output bigger for real instances than for fake instances. Therefore, the discriminator in a WGAN is usually referred to as a critic rather than a discriminator. The discriminator tries to maximize the critic loss $D(x) - D(G(z))$, where $D(x)$ is the critic's output for a real instance, $G(z)$ is the generator's output for given z. $D(G(z))$ is the critic's output for fake data. Thus, it tries to maximize the difference between its output on real data and its output on fake data. The generator tries to maximize the generator loss $D(G(z))$. Thus, it tries to maximize the discriminator's output for its fake data. WGANs are less vulnerable to suffering model collapse and can avoid vanishing gradients issues.

1.6.6 Transformer

Transformer has been one of the most popular deep learning architectures recently due to its usability in a wide range of applications from natural language processing tasks to visual tasks and its state-of-the-art results in multiple public datasets. However, it is important to note that transformers come with high computational and memory requirements that might not be ideal for embedded solutions. Some works like Attention Is All You Need [44] focus on tackling the mentioned bottlenecks while mimicking the functions of a transformer. In the following paragraph, we provide an explanation of different blocks in a transformer which might ease readers to understand related works.

In [44] is introduced the idea of a transformer which is made of six encoders, six decoders, and uses machine translation as an application. A machine translation task takes a sentence or a phrase (sequence of words) as input and outputs the phrase translated into the target language. Each encoder is identical in architecture while having their own set of learnable weights and consists of a self-attention and a feedforward layer. The self-attention layer can be regarded as a context-aware encoding mechanism where it uses information from other words to encode it better. In a technical implementation perspective, the self-attention mechanism involves for each word three vectors namely Attention Is All You Need [44] query (Q), key (K), and value (V) which are generated by three different fully connected layers having output dimension smaller than that of the input embedding vector. In order to compute a score for each word against all other words in the phrase, a dot product is taken between the query vector of the word and key vector of all the words in the phrase. The score is further divided by the square root of the dimension of the key vector d_K to stabilize the training by normalizing the gradients. Next, all the scores are passed through a softmax function to generate a normalized distribution. Finally, as shown in Eq. (1.56), the softmax output is multiplied with the value vector matrix to generate the output Z of the self-attention layer for the given position. This output is then simply fed to the following fully connected layer:

$$Z = \text{softmax}\left(\frac{Q \times K^T}{\sqrt{d_K}}\right) V \tag{1.56}$$

Transformers also introduce the idea of multihead attention that involves having multiple self-attention layers initialized randomly in order to have different encodings to cover multiple subspaces. The multihead attention generates multiple outputs Z which are concatenated together and multiplied with a jointly trained weight vector W that projects them into a single vector which is fed to the

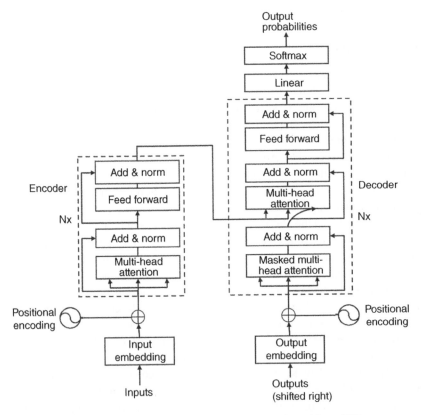

Figure 1.17 Transformer network architecture. Source: adapted from [44].

fully connected layer. In order to capture the order of words in a given sequence, a positional encoding is generated where each element in the encoding represent a sinusoid. This is then added to the word-embedding vector which results in the input vector to the encoder. Additionally, each encoder consists of a residual connection with a normalization layer. In the decoder, the incoming key and value vectors coming from the top encoder act as input which is used by the encoder–decoder attention layer. After each time step, the output of the decoder is fed back to the decoder along with a positional encoding which becomes the input for the self-attention layer in the decoder. The decoder self-attention layer is prevented from attending positions in future by masking all remaining places which are yet to be predicted to $-\infty$ and only the predicted output sequence goes as input to the self-attention layer in the decoder. The final layer of the decoder consists of a logit layer which has the dimension of all the possible words, and softmax is applied on it to select the one with highest probability as the predicted word.

The overall architecture of a transformer is depicted in Figure 1.17.

1.7 Training a Neural Network

For training a neural network, there are two steps namely forward pass and error backpropagation.

1.7.1 Forward Pass and Backpropagation

In the forward pass, the input is fed to the model and multiplied with weight vectors, and bias is added for each layer to compute the output of the model. The input x_i^l, activation u_j^l, and output o_j^l at lth layer of dense or fully connected layer is represented as follows:

$$x_i^{(l)} = o_j(l-1)$$
$$u_j^{(l)} = \sum_{i=1}^{N} w_{ij}^{(l)} x_i^{(l)}$$
$$o_j^{(l)} = \sigma(u_j^{(l)}) \tag{1.57}$$

where N denotes the number of neurons at the lth layer, w_{ij}^l are the weights that need to be learned for a task at the lth layer and $\sigma()$ is the activation function.

The backpropagation is described below. Considering one sample for which inputs $(x_1, x_2, \ldots, x_n)$ and expected outputs $(t_1, t_2, \ldots t_k, \ldots, t_m)$ and real outputs $(y_1, y_2, \ldots y_k, \ldots, y_m)$, the error for one sample is therefore $E = \frac{1}{2} \sum_{k=1}^{m} (y_k - t_k)^2$, where y_k is a function of the weights $w_{ij}^{(l)}$. The weights are updated to minimize the error using the gradient descent algorithm and can be expressed as follows:

$$w_{ij}^{(l)} \leftarrow w_{ij}^{(l)} - \lambda \frac{\partial E}{\partial w_{ij}^{(l)}} \tag{1.58}$$

In Eq. (1.58), $\frac{\partial E}{\partial w_{ij}^{(l)}}$ can be computed as follows:

$$\frac{\partial E}{\partial w_{ij}^{(l)}} = \frac{1}{2} \sum_k \frac{\partial E}{\partial y_k} \frac{\partial y_k}{\partial w_{ij}^{(l)}} \tag{1.59}$$

where $\frac{\partial E}{\partial y_k} = (y_k - t_k)$.

Since y_k is a function of $u_j^{(l)}$, it can be derived that

$$\frac{\partial y_k}{\partial w_{ij}^{(l)}} = \frac{\partial y_k}{\partial u_j^{(l)}} \frac{\partial u_j^{(l)}}{\partial w_{ij}^{(l)}} \tag{1.60}$$

$$\frac{\partial u_j^{(l)}}{\partial w_{ij}^{(l)}} = o_i^{(l-1)} \tag{1.61}$$

which is computed during the feed forward step.

Thus, putting it all together gives us

$$\frac{\partial E}{\partial w_{ij}^{(l)}} = (y_j - t_j) \frac{\partial y_k}{\partial u_j^{(l)}} o_i^{(l-1)} = \frac{\partial E}{\partial u_j^{(l)}} o_i^{(l-1)} \tag{1.62}$$

Some of the important aspects during training of a neural network are the following:

1. Learning Rate: Each weight update is controlled by parameter λ known as the learning rate parameter. If the learning rate is too small, then it may result in very slow learning, can get trapped in local minima easily, and can keep running for many iterations. On the other hand, if the learning rate is large, then it may step over the minima, can fail to converge, and potentially diverge. So it is really important to choose a good learning rate based on the architecture, dataset, transfer function, etc. Figure 1.18 illustrates the effects of choosing small and large learning rates on the gradient descent.

2. Weight Initialization: It is important to randomize the weights during initialization; otherwise, symmetry in weights would prevent the network from learning. Usually, small random values are used which is highly important when the number of neurons in a layer grows, as the weighted sum may saturate the optimization function.

3. Overfitting and Underfitting: In machine learning, the objective is not only to minimize cost function on in-sample data, i.e., data available or seen, but also generalize on out-sample data, i.e., data not available or unseen during training.

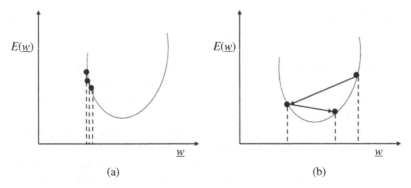

Figure 1.18 Illustration of the gradient descent when (a) the learning rate is small, and (b) the learning rate is large.

During training, the available dataset is divided into a training set, a validation set, and a test set. The training dataset is used to train the model, the validation dataset is used to set the hyperparameters of the model, and the test dataset is used for estimating the out-sample or generalization accuracy.

When the performance is poor on the training data, then it can be regarded as underfitting and is often due to poor choice of learning rate or if the neural network is under-dimensional. This error is referred to as "bias." The issue of underfitting is illustrated in the left column of Figure 1.19. The issue of overfitting arises when the performance is good on the training data, i.e., good approximation accuracy, but poor on the test or validation data, i.e., poor generalization accuracy. This phenomenon is also referred as "variance" and is illustrated in the right column of Figure 1.19. If the training set size is insufficient or the model complexity is too high for the data, the model memorizes or approximates the training data well but does not generalize well on test data, i.e., it

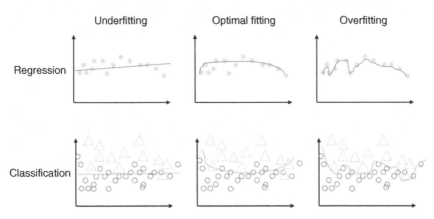

Figure 1.19 Illustration of underfitting and overfitting of a model.

overfits. The purpose of training a machine-learning model is to find a model as shown in the middle column of Figure 1.19, where the training error (bias) as well as the generalization error (variance) is minimized. Typically, training finds a model such that a balance between bias and variance can be achieved and often is referred as "bias-variance" trade-off. In case of deep learning, the "bias-variance" trade-off is not applicable since there are separate mechanisms to reduce bias and variance, thus the trade-off is not readily applicable.

4. Curse of Dimensionality: The other critical aspect in machine learning in general is the curse of dimensionality. Curse of dimensionality is closely related to overfitting. In high-dimensional spaces, most of the training data resides in the corners of the hypercube defining the feature space. Instances in the corners of the feature space are much more difficult to classify than instances around the centroid of the hyperactive sphere. Thus, as the number of features or dimensions grows, the amount of data we need to generalize accurately also grows exponentially.

1.7.2 Optimizers

Optimizers are methods that help to change weights and bias of the model such that a loss function is minimized. There are several modifications that have been proposed to the standard stochastic gradient descent (SGD) algorithm that $w_{t+1} = w_t - \lambda \partial_w E(w)$, where $E(w), \partial E(w)$ denotes the loss function and its derivative, respectively. w_{t+1} and w_t represents the weights after and before the update step and λ represents the learning rate. Following are list of optimizers that have been proposed in the that improves the standard SGD:

1. Momentum: It accelerates the SGD toward the relevant direction while reducing the oscillations. It basically adds a part of the previous weight updates to the current update vector ensuring that the direction of the previous update is retained to some extent while the current update gradient is used to fine-tune the final update direction. Momentum introduces another variable v_t and can be expressed as follows:

$$v_t = \gamma v_{t-1} + \lambda \partial_w E(w)$$
$$w_{t+1} = w_t - v_t \tag{1.63}$$

2. Nesterov Accelerated Gradient [45]: While momentum helps to reduce noise and also accelerates the convergence, it also introduces error. This is resolved in Nesterov accelerated gradient by including part of the previous weight updates to the current update vector to perform the weight update that is expressed as follows:

$$v_t = \gamma v_{t-1} + \lambda \partial_w E(w - \gamma v_{t-1})$$
$$w_{t+1} = w_t - v_t \tag{1.64}$$

A typical value of $\gamma = 0.9$.

3. Adagrad [46]: The motivation of Adagrad is to have an adaptive learning rate for each parameter; however, earlier approaches have fixed learning rate. Adagrad ensures that different neurons of the hidden layer dependent on iterations have different learning rates. The intuition behind it is that large updates should occur for infrequent parameters and smaller for frequent parameters.

For each weight updates, the learning rate is adapted as follows:

$$w_{t+1}t^i = w_t^i - \frac{\lambda}{\sqrt{G_t^i + \epsilon}} \partial_w E(w_t^i)$$

$$G_t^i = \sum_{tt=0}^{t} \left(\partial_w E(w_{tt}^i) \right)^2 \tag{1.65}$$

since the sum of squared gradients grows continuously, it would lead to a smaller learning rates adaptively. The parameter ϵ helps in avoiding divide by zero issues.

4. RMSprop [47]: An issue of Adagrad is that after several iterations in DNNs, the learning rate becomes very small leading to the issue of dead neuron problems and results in no updates for these neurons. This issue is fixed by RMSprop, where learning can continue even after many parameter updates. In RMSprop, the learning rate is an exponential average of the gradients instead of the cumulative sum of squared gradients as in Adagrad. A moving average of a squared gradient for each weight is computed by limiting the gradient accumulation to a certain past and can be expressed as follows:

$$w_{t+1}^i = w_t^i - \frac{\lambda}{\sqrt{G_t^i + \epsilon}} \partial_w E(w_t^i)$$

$$G_t^i = \gamma G_{t-1}^i + (1 - \gamma) \left(\partial_w E(w_t^i) \right)^2 \tag{1.66}$$

5. Adadelta [48]: Adadelta is another improvement over the Adagrad to continue learning after many parameter updates. But Adadelta is computationally expensive. Here the gradient accumulation is limited to a certain past update by computing a moving average of both the squared gradient and parameter updates for each weight parameter as follows:

$$w_{t+1}^i = w_t^i - \lambda_t \partial_w E(w_t^i) = w_t^i + v_t^i$$

$$G_{t+1}^i = \gamma G_t^i + (1 - \gamma) \left(\partial_w E(w_{t-1}^i) \right)^2$$

$$\Delta w_{t+1}^i = \gamma \Delta w_t^i + (1 - \gamma) \left(v_{t-1}^i \right)^2$$

$$\lambda_t = \frac{\sqrt{\Delta w_t^i + \epsilon}}{\sqrt{G_t^i + \epsilon}} \tag{1.67}$$

6. Adaptive Moment Estimation (ADAM) [49]: Adam optimizer is one of the most popular and widely used one today. It stores both the decaying average of the

past gradients similar to momentum and also the decaying average of the past squared gradients, similar to RMSprop and Adadelta. ADAM can be expressed as in the following equation, where momentum is added to RMSprop by using first and second moments, i.e., mean m_{t+1}^i and variance v_{t+1}^i of the gradient:

$$m_{t+1}^i = \beta_1 m_t^i + (1 - \beta_1)\partial_w E(w_t^i)$$

$$v_{t+1}^i = \beta_2 v_t^i + (1 - \beta_2)\big(\partial_w E(w_t^i)\big)^2$$

$$w_t^i = w_{t-1}^i - \frac{\lambda}{\sqrt{v_t^i + \epsilon}} m_t^i \qquad (1.68)$$

where β_1 and β_2 are the forgetting factor in the moving average implementation of the mean and variance of the gradient. Adam is easy to implement and computationally efficient and requires less memory owing to the moving average implementation.

1.7.3 Loss Functions

A neural network is formulated as an optimization problem. The candidate solution, which means the weights of the network, should minimize or maximize the score of the given objective function.

In the case of a regression problem, the objective is to predict a real-value quantity. In this case, linear activation unit is used at the output layer, and MSE is used as the loss function. The mean-square loss for regression is given as follows:

$$\mathrm{MSE}(y, \hat{y}) = |y - \hat{y}|^2 \qquad (1.69)$$

where y and $\hat{y}$ are the true value and predicted value of the neural network, respectively.

For modeling a classification problem, the idea is to map the input variable to a class label implying that the objective is to predict the probability of an example for belonging to a particular class. Under maximum likelihood estimation, the training of the network is seeking to find a set of model weights that minimizes the difference between the model's predicted probability distribution given the dataset and the distribution of probabilities in the training dataset. This is called the CE loss, and in the case of binary classification is configured as a sigmoid activation at the output, while for multiclass classification, the softmax activation is used at the output. In both cases, the problem is formulated as predicting the maximum likelihood for a given input belonging to a particular class.

The binary CE loss for binary classification is given as follows:

$$\mathrm{CE}(p, \hat{y}) = -\hat{y} \log(p) - (1 - \hat{y}) \log(1 - p) \qquad (1.70)$$

where p is the probability of class 1, $1 - p$ is the probability of class 0, and $\hat{y}$ is the predicted probability from the neural network.

1.8 Questions to the Reader

- Explain the processing pipeline of 2D angle-Doppler image-based detection. Explain a use-case where 2D image-Doppler image is preferred over range-Doppler or range-angle images.
- What are the adaptation in processing pipeline for sensing extended targets compared to point target processing?
- What are the advantages of OS-CFAR detector over CA-CFAR detector for extended targets? What is the need for guard cells in CA-CFAR? And why are they not absolutely necessary for OS-CFAR?
- What are the advantages and disadvantages of DBSCAN and Euclidean clustering?
- What does Cramer–Rao bound for estimating a radar parameter indicate? Derive the Cramer–Rao bound for angle estimation.
- What is the purpose of introducing nonlinearities in neural networks? How is it achieved in convolutional neural network or LSTM?
- Why is initializing all the weights of a neural network with the same value (e.g., 0.1) not a good idea?
- How does 2D CNN ensure invariance toward spatial translation? How can you extend 2D CNN to 3D CNN and which application can you think of?
- Explain the different configuration LSTM that can be used in a radar task, e.g., people counting or gesture sensing.
- Explain the reparameterization trick in VAE. Explain the advantage of GAN loss over VAE loss.
- What is mode collapse in GAN? How to identify a stable GAN training?
- What are the improvements proposed by Wasserstein GAN? How is the Lipschutz continuity implemented in practical Wasserstein GAN.
- What is bias–variance trade-off? What are the means of dealing with bias and variance in a neural network?
- What is global receptive field in deep CNN (DCNN)?
- What is the issue of Adagrad optimizer that RMSprop solves? What is the problem of dead neurons?

References

1 Meinel, H.H. (2014). Evolving automotive radar - from the very beginnings into the future. *The 8th European Conference on Antennas and Propagation (EuCAP 2014)*, pp. 3107–3114.
2 Li, Y., Du, L., and Liu, H. (2011). Moving vehicle classification based on micro-Doppler signature. *2011 IEEE International Conference on Signal*

Processing, Communications and Computing (ICSPCC), September 2011, pp. 1–4.

3 Dubey, A., Fuchs, J., Reissland, T. et al. (2020). Uncertainty analysis of deep neural network for classification of vulnerable road users using micro-Doppler. *2020 IEEE Topical Conference on Wireless Sensors and Sensor Networks (WiSNeT)*, pp. 23–26.

4 Xu, R. and Wunsch, D. (2005). Survey of clustering algorithms. *IEEE Transactions on Neural Networks* 16 (3): 645–678.

5 Ester, M., Kriegel, H.-P., Sander, J., and Xu, X. (1996). A density-based algorithm for discovering clusters in large spatial databases with noise. *KDD-96* 96 (34): 226–231.

6 Dubey, A., Fuchs, J., Luebke, M. et al. (2020). Generative adversial network based extended target detection for automotive MIMO radar. *2020 IEEE International Radar Conference (RADAR)*, pp. 220–225.

7 Stephan, M. and Santra, A. (2019). Radar-based human target detection using deep residual U-net for smart home applications. *2019 18th IEEE International Conference On Machine Learning And Applications (ICMLA)*, pp. 175–182.

8 Dubey, A., Fuchs, J., Luebke, M. et al. (2020). Region based single-stage interference mitigation and target detection. *IEEE Radar Conference 2020*.

9 Qi, C., Su, H., Mo, K., and Guibas, L.J. (2016). PointNet: Deep learning on point sets for 3D classification and segmentation. *CoRR*, vol. abs/1612.00593. http://arxiv.org/abs/1612.00593.

10 Qi, C.R., Yi, L., Su, H., and Guibas, L.J. (2017). PointNet++: Deep hierarchical feature learning on point sets in a metric space. *CoRR*, vol. abs/1706.02413. http://arxiv.org/abs/1706.02413.

11 Guo, Y., Wang, H., Hu, Q. et al. (2019). Deep learning for 3D point clouds: a survey. *CoRR*, vol. abs/1912.12033. http://arxiv.org/abs/1912.12033.

12 Danzer, A., Griebel, T., Bach, M., and Dietmayer, K. (2019). 2D car detection in radar data with pointnets. *CoRR*, vol. abs/1904.08414. http://arxiv.org/abs/1904.08414.

13 Lee, S. (2020). Deep learning on radar centric 3D object detection.

14 Qi, C.R., Liu, W., Wu, C. et al. (2017). Frustum pointnets for 3D object detection from RGB-D data. *CoRR*, vol. abs/1711.08488. http://arxiv.org/abs/1711.08488.

15 Yang, Z., Sun, Y., Liu, S., and Jia, J. (2020). 3DSSD: Point-based 3D single stage object detector. *CoRR*, vol. abs/2002.10187. https://arxiv.org/abs/2002.10187.

16 Simon, M., Milz, S., Amende, K., and Gross, H. (2018). Complex-YOLO: Real-time 3D object detection on point clouds. *CoRR*, vol. abs/1803.06199. http://arxiv.org/abs/1803.06199.

17 Shi, S., Wang, Z., Wang, X., and Li, H. (2019). Part-a^2 net: 3D part-aware and aggregation neural network for object detection from point cloud. *CoRR*, vol. abs/1907.03670. http://arxiv.org/abs/1907.03670.

18 Yang, B., Wang, J., Clark, R. et al. (2019). Learning object bounding boxes for 3D instance segmentation on point clouds. *CoRR*, vol. abs/1906.01140. http://arxiv.org/abs/1906.01140.

19 Jiang, M., Wu, Y., and Lu, C. (2018). PointSIFT: A sift-like network module for 3D point cloud semantic segmentation. *CoRR*, vol. abs/1807.00652. http://arxiv.org/abs/1807.00652.

20 Duan, Z., Li, X.R., Han, C., and Zhu, H. (2005). Sequential unscented Kalman filter for radar target tracking with range rate measurements. *2005 7th International Conference on Information Fusion*, Volume 1, p. 8.

21 Lin, A. and Ling, H. (2006). Three-dimensional tracking of humans using very low-complexity radar. *Electronics Letters* 42 (18): 1062–1063.

22 Chang, S., Wolf, M., and Burdick, J.W. (2009). An MHT algorithm for UWB radar-based multiple human target tracking. *2009 IEEE International Conference on Ultra-Wideband*, pp. 459–463.

23 Kim, Y. and Ling, H. (2009). Through-wall human tracking with multiple Doppler sensors using an artificial neural network. *IEEE Transactions on Antennas and Propagation* 57 (7): 2116–2122.

24 Cortina, E., Otero, D., and D'Attellis, C.E. (1991). Maneuvering target tracking using extended Kalman filter. *IEEE Transactions on Aerospace and Electronic Systems* 27 (1): 155–158.

25 Chang, S., Sharan, R., Wolf, M. et al. (2009). UWB radar-based human target tracking. *2009 IEEE Radar Conference*, pp. 1–6.

26 Chang, S., Wolf, M., and Burdick, J.W. (2010). Human detection and tracking via ultra-wideband (UWB) radar. *2010 IEEE International Conference on Robotics and Automation*, pp. 452–457.

27 Sun, W., Huang, W., Ji, Y. et al. (2019). A vessel azimuth and course joint re-estimation method for compact HFSWR. *IEEE Transactions on Geoscience and Remote Sensing* 10: 1–11.

28 Vaishnav, P. and Santra, A. (2020). Continuous human activity classification with unscented Kalman filter tracking using FMCW radar *IEEE Sensors Letters* 4 (5): 1–4.

29 Redmon, J., Divvala, S., Girshick, R., and Farhadi, A. (2016). You only look once: unified, real-time object detection. *Proceedings of the IEEE Conference on Computer Vision and Pattern Recognition*, pp. 779–788.

30 Liu, W., Anguelov, D., Erhan, D. et al. (2016). SSD: Single shot multibox detector. *European Conference on Computer Vision*, Springer, pp. 21–37.

31 Dai, J., Li, Y., He, K., and Sun, J. (2016). R-FCN: Object detection via region-based fully convolutional networks. In: *Advances in Neural Information Processing Systems*, vol. 29.

32 Prophet, R., Hoffmann, M., Vossiek, M. et al. (2018). Pedestrian classification with a 79 GHz automotive radar sensor. *2018 19th International Radar Symposium (IRS)*, pp. 1–6.

33 Prophet, R., Hoffmann, M., Ossowska, A. et al. (2018). Pedestrian classification for 79 GHz automotive radar systems. *2018 IEEE Intelligent Vehicles Symposium (IV)*, pp. 1265–1270.

34 Prophet, R., Hoffmann, M., Ossowska, A. et al. (2018). Image-based pedestrian classification for 79 GHz automotive radar. *2018 15th European Radar Conference (EuRAD)*, pp. 75–78.

35 LeCun, Y. and Bengio, Y. (1995). Convolutional networks for images, speech, and time series. *The Handbook of Brain Theory and Neural Networks* 3361 (10): 1995.

36 Krizhevsky, A., Sutskever, I., and Hinton, G.E. (2012). ImageNet classification with deep convolutional neural networks. In: *Advances in Neural Information Processing Systems*, vol. 25.

37 Szegedy, C., Ioffe, S., Vanhoucke, V., and Alemi, A.A. (2017). Inception-v4, inception-ResNet and the impact of residual connections on learning. *31st AAAI Conference on Artificial Intelligence*.

38 Hochreiter, S. and Schmidhuber, J. (1997). Long short-term memory. *Neural Computation* 9 (8): 1735–1780.

39 Chung, J., Gulcehre, C., Cho, K., and Bengio, Y. (2014). Empirical evaluation of gated recurrent neural networks on sequence modeling. *arXiv preprint arXiv:1412.3555*.

40 Kingma, D.P. and Welling, M. (2019). An introduction to variational autoencoders. *Foundations and Trends® in Machine Learning* 12 (4): 307–392.

41 Kipf, T.N. and Welling, M. (2016). Variational graph auto-encoders. *arXiv preprint arXiv:1611.07308*.

42 Pu, Y., Gan, Z., Henao, R. et al. (2016). Variational autoencoder for deep learning of images, labels and captions. In: *Advances in Neural Information Processing Systems*, vol. 29.

43 Goodfellow, I., Pouget-Abadie, J., Mirza, M. et al. (2014). Generative adversarial nets. In: *Advances in Neural Information Processing Systems*, vol. 27.

44 Vaswani, A., Shazeer, N., Parmar, N. et al. (2017). Attention is all you need. In: *Advances in Neural Information Processing Systems*, vol. 30.

45 Nesterov, Y. (2013). Gradient methods for minimizing composite functions. *Mathematical Programming* 140 (1): 125–161.

46 Duchi, J., Hazan, E., and Singer, Y. (2011). Adaptive subgradient methods for online learning and stochastic optimization. *Journal of Machine Learning Research* 12 (7).

47 Tieleman, T. and Hinton, G. (2012). Lecture 6.5-Rmsprop, Coursera: Neural Networks for Machine Learning. *University of Toronto, Technical Report*, vol. 6.

48 Zeiler, M.D. (2012). ADADELTA: An adaptive learning rate method. *arXiv preprint arXiv:1212.5701*.

49 Kingma, D.P. and Ba, J. (2014). Adam: A method for stochastic optimization. *arXiv preprint arXiv:1412.6980*.

2

Deep Metric Learning

After reading this chapter, you should be able to

- Understand the basic concept of deep metric learning.
- Have an overview of the state-of-the-art deep metric learning techniques.
- Understand how deep metric learning helps to improve real-world radar applications.

There are several radar applications that aim to classify between a set of predefined classes such as different human activities or hand gestures. However, in real-world environments, more than the predefined classes exist, which turns the problem into an open-set classification task. An open-set classification means that the network should be able to detect if an input belongs to one of the predefined or known classes or not. For radar-based hand gesture recognition, this might be a random body movement or a hand movement not intended to be a gesture such as scratching the nose or reaching for a glass of water. Conventional deep-learning classifiers use a fully connected layer with softmax activation as final layer and train the network using the cross-entropy loss. The softmax activation maps the class scores to a probability distribution over the known classes. Consequently, the probabilities of the known classes sum up to one. A probability for being none of the predefined classes is not considered, which shows the closed-set assumption of softmax. This results in a good classification accuracy on the set of known classes but performs very poorly in detecting an unknown motion input.

An approach to tackle this issue is to introduce a garbage class. This requires training samples of motions that should not be classified as one of the known classes. However, it is almost infeasible or at least connected with a very large effort to record a garbage data set that represents all possible motions that might appear in real-world environments. It is much more desirable to train the network only with the known classes in a way that it is additionally able to detect when an input is not within the set of known classes. To achieve this, the network has to learn

Methods and Techniques in Deep Learning: Advancements in mmWave Radar Solutions, First Edition.
Avik Santra, Souvik Hazra, Lorenzo Servadei, Thomas Stadelmayer, Michael Stephan, and Anand Dubey.
© 2023 The Institute of Electrical and Electronics Engineers, Inc. Published 2023 by John Wiley & Sons, Inc.

very class-specific and discriminative class features. One well-known approach to achieve this is metric learning. Deep metric learning aims to learn a relation or similarity between the training samples. Similar samples, which are samples of the same class, are pulled together, whereas dissimilar samples, which are samples of two different classes, are pushed apart. By this learning concept, the features are clustered per class in the feature space. After training the class clusters are compact and far apart. Therefore, it arises as a gap between the class clusters. If an input sample is projected into this gap, it can be detected as an outlier and thus as an unknown input class.

The remaining chapter is structured as follows: first an introduction to metric learning is given, and then a general understanding and overview of the most important metric learning techniques is provided. The taxonomy as proposed in Figure 2.1 is used throughout this chapter. We divide the metric learning approaches into three main categories, which are the pairwise methods, the proxy methods, and the end-to-end methods. Pairwise methods learn a metric based on similarities between different training samples and require postprocessing for final classification. In the proxy methods, the metric is learned by evaluating the similarity between a training sample and the corresponding class-representative proxy vector, and in the end-to-end methods, the learning of the metric is integrated into the neural network and the classification results are output directly. Each of the three main approaches can work with different concepts, that are

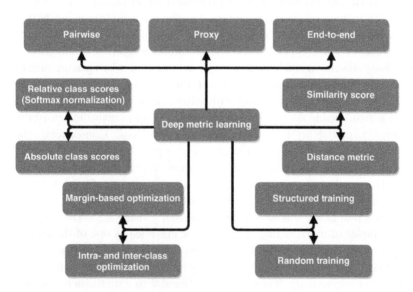

Figure 2.1 Taxonomy of the deep metric learning approaches.

depicted in Figure 2.1, such as using a margin-based optimization or an intra- and interclass optimization. The used terms in Figure 2.1 are introduced and explained throughout the chapter. At the end of the chapter, a radar-based gesture sensing application and the positive impact of metric learning is presented.

2.1   Introduction

The aim of metric learning is to learn a relation or similarity between samples in a data set. A very common metric to learn is the Mahalanobis distance, because it has, unlike many other metrics such as the Euclidean distance, tunable parameters. The Mahalanobis distance is defined as follows:

$$d_{\mathrm{M}}(x_1, x_2) = \sqrt{(x_1 - x_2)^T M (x_1 - x_2)} \tag{2.1}$$

where M is a positive definite matrix. The entries of the matrix M can be optimized according to a specific objective. Since M is symmetric and positive definite, it can be split up into $M = L^T L$, and the Mahalanobis distance reformulated as follows:

$$\begin{aligned} d_{\mathrm{M}}(x_1, x_2) &= \sqrt{(x_1 - x_2)^T M (x_1 - x_2)} \\ &= \sqrt{(x_1 - x_2)^T L^T L (x_1 - x_2)} \\ &= \sqrt{(Lx_1 - Lx_2)^T (Lx_1 - Lx_2)} \\ &= \| Lx_1 - Lx_2 \|_2 \end{aligned} \tag{2.2}$$

which means that it is equal to the Euclidean distance of the linearly transformed input vectors x_1 and x_2. For deep metric learning, this linear transformation is replaced by a nonlinear mapping $f(x)$ performed by the neural network as stated in Eq. (2.3). Thus, for deep metric learning, the goal is to tune the parameter of the entire neural network in a way that a simple distance metric such as the Euclidean distance describes the similarity of the data well.

$$d_{\mathrm{deep}}(x_1, x_2) = \| f(x_1) - f(x_2) \|_2 \tag{2.3}$$

In this book, we are focusing on supervised deep metric learning for classification tasks. Hence, the similarity gets discrete. Similar samples are samples of the same class, and dissimilar samples are ones of different classes. Thus, the objective in the discrete world of a supervised classification problem is that the distance of samples to their true class center should be significantly smaller than to other class centers. Ideally, there should be a margin between the classes. In a wide sense, any approach whose loss function is given by constraints on distances or similarities can be considered as metric learning [1].

2.2 Pairwise Methods

The pairwise metric learning approaches are learning similarities between the samples itself. Therefore, the respective loss functions always take two or more samples into account (Figure 2.2).

2.2.1 Contrastive Loss

One of the first deep metric learning approaches was presented in 2005 by Chopra et al. [2]. The loss is calculated on a pair of samples. The samples within a pair are either from the same class or from different classes. If the samples are from the same class, then the objective is to minimize the distance between both feature vectors; however, if the samples are from different classes, the objective is to maximize the distance between their respective feature vectors. The contrastive loss function is defined as follows:

$$L_{\text{cont}}(x_i, x_j) = \begin{cases} \|f(x_i) - f(x_j)\|_2^2, & \text{if } y_i = y_j \\ \max\left(0, m - \|f(x_i) - f(x_j)\|_2^2\right), & \text{otherwise} \end{cases} \quad (2.4)$$

where m is a hyperparameter introducing a minimal required margin between two dissimilar samples.

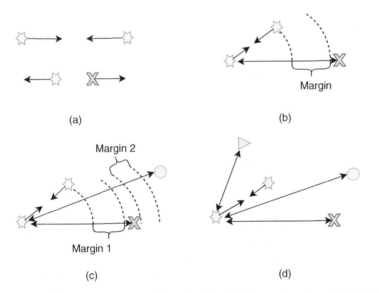

(a)

(b)

(c)

(d)

Figure 2.2 Overview of metric learning losses between samples. (a) Contrastive loss, (b) triplet loss, (c) quadruplet loss, and (d) *N*-pair loss.

2.2.2 Triplet Loss

The triplet loss is very related to the aforementioned contrastive loss. The idea of triplet learning was first published by Schroff et al. in 2015 [3]. The objective remains the same. Minimize the distance between samples of the same class and push away samples of other classes more than a certain threshold. The difference is that the loss function is evaluated on a triplet instead of a pair of samples. The triplet is composed of an anchor, a positive, and a negative sample. The anchor is the actual training sample, the positive one is another sample of the same class, and the negative one is a sample from any other class. The loss function is defined as follows:

$$L_{\text{triplet}}(x_a, x_p, x_n) = \max \left(\|f(x_a) - f(x_p)\|_2^2 + m - \|f(x_a) - f(x_n)\|_2^2 \right) \qquad (2.5)$$

where m is the margin that is at least required between the anchor and negative samples. The triplet learning approach is highly dependent on the choice of good positive and negative examples.

2.2.3 Quadruplet Loss

The quadruple loss [4] extends the triplet loss by another negative sample. The additional negative sample is not only from another class as the anchor but also from another class as the negative sample. The quadruplet loss uses the distance between anchor and positive $d(x_a, x_p)$, anchor and negative $d(x_a, x_n)$, and negative and negative $d(x_n, x_m)$. The first part of the quadruplet loss is exactly the same as the triplet loss using $d(x_a, x_p)$ and $d(x_a, x_n)$. However, the idea of the quadruplet loss is to add an auxiliary task, which is that the distance between the two negative samples has to be also larger than the distance between anchor and positive. The distance between anchor and positive $d(x_a, x_p)$ represent an intraclass distance, whereas the distance between negative pairs, either $d(x_a, x_n)$ or $d(x_a, x_m)$, represent an interclass distance. By the additional loss term, it is not only ensured that the intraclass distance between the positive pair is smaller than the distance to a negative sample, but also that the intraclass distance is smaller than the interclass distance between any arbitrary pair of classes. The loss objective is formulated as follows:

$$L_{\text{quad}}(x_a, x_p, x_n, x_m) = \max \left(\|f(x_a) - f(x_p)\|_2^2 + m_1 - \|f(x_a) - f(x_n)\|_2^2 \right)$$
$$+ \max \left(\|f(x_a) - f(x_p)\|_2^2 + m_2 - \|f(x_n) - f(x_m)\|_2^2 \right)$$
$$(2.6)$$

where m_1 and m_1 are two independent margin parameters. The authors of [4] point out that the right order of the samples is still obtained by the triplet part in the loss function. The additional term is an auxiliary task that can help to further increase the interclass distances and increase the performance on the test data. Therefore, the auxiliary loss term should not have the same impact on the training as the triplet loss part. Thus, the authors propose to choose margins so that $m_2 \langle m_1$.

2.2.4 N-Pair Loss

The N-pair loss [5] goes even one step further than the quadruplet loss and takes multiple negative pairs into account. The reasoning behind this is that in the triplet loss, where only one positive and one negative pair is used, the single negative pair insufficiently represents all negative classes. In other words, there is only one intraclass distance, that is represented in each triplet, but many possible interclass distances, since there are many different classes, where only one out of many possibilities is represented. If by chance the randomly selected interclass distance is large, then the triplet loss is satisfied, although there might be many examples from other negative classes that are much too close. The hope of triplet loss is that over a large number of triplets, all classes are finally represented. However, this requires a very large number of triplets and still the imbalance of intraclass distance compared to each time a different interclass distance consists. The loss objective of the N-pair loss is given as follows:

$$
\begin{aligned}
L_{\text{N--pair}}(x_a, x_p, x_1, x_2, ..., x_{N-1}) &= \log \left(1 + \sum_{i=1}^{N-1} \exp(x_a^T x_i - x_a^T x_p) \right) \\
&= -\log \left(\frac{\exp(x_a x_p)}{\exp(x_a x_p) + \sum_{i=1}^{N-1} \exp(x_a^T x_i)} \right)
\end{aligned}
\tag{2.7}
$$

Please note that this is the softmax loss for multiclass classification that is introduced in Section 2.3 in Eq. (2.13).

2.2.5 Summary

Additionally to the presented basic pairwise metric learning methods, there exist many extensions and variations in literature such as the magnet loss [6], the structured loss [7], the clustering loss [8], the mixed loss [9], or the multisimilarity loss [10] to name some of them. A detailed overview of these methods can be found in [11]. The pairwise methods show very good results and have attracted a lot of attention. However, they also encounter some problems. A single pair or triplet represents the data very poorly, which is why over time, it was proposed to take more and more samples into account. The first approach was working on pairs, then Schroff et al. introduced the triplets, after that quadruplets where proposed, and finally, the N-pair loss proposes to take even N samples into account. With more samples, the data are better represented; however, it opens another problem. The combinatorial possibilities are exploding by increasing number of samples. Additionally, many combinations are poor training pairs or triplets. This is why even strategies for triplet generation and finding hard negative triplets were proposed.

Furthermore, the pairwise approaches do not learn a classification per se, but a similarity within the data. Of course, the definition of similarity is predefined as a supervised information. If we define similarity with respect to a classification task, then the similarity gets discrete according to the classes. Samples belonging to the same class are similar, and samples from different classes are dissimilar. There is nothing in between. Due to this discrete similarity, the network learns to project the embedded features to well separated and compact class clusters. However, no class scores are predicted by the network, and therefore, a postprocessing is required. The network was trained in a way that the Euclidean distance represents well the similarity of the data samples. Therefore, a very common method to extend the trained network to a classifier is the K-nearest neighbor (K-NN) classifier. It assigns the test sample to the dominant class within its K-nearest neighbors.

Please note that we are in this book discussing the pairwise metric learning approaches from a classification perspective. However, we would like to point out that there are also different applications where these methods show great results such as image-retrieval tasks [12–14].

2.3 End-to-End Learning

The pairwise methods learn a similarity between the samples. Since for the classification use case the similarity is defined by the discrete class labels, the samples are implicitly grouped in the embedding space by classes. However, the network does not output class probabilities as it is the case for conventional deep learning classifiers using a softmax layer; therefore, an additional clustering algorithm such as K-NN is needed to do the class assignment.

As already mentioned conventional deep neural network classifiers use a softmax activation as an output layer. The network is then trained based on the cross-entropy between the output vector of the neural network and an one-hot encoded label vector. This approach provides directly class scores and simplifies the class assignment. However, the problem is that the closed-set assumption of the softmax maps the class scores to a probability distribution of only the known classes, and no unknown class is considered. To solve this issue, a thresholding can be applied on the class scores. If there is no class with a probability higher than a certain threshold, then the input is rejected. However, also unknown samples are often assigned to one of the known classes with a very high probability, which leads to bad rejection accuracy using simple thresholding [15]. Another naive approach to solve this problem is to introduce a garbage class and record a training data set including alien motions to better reflect the real world. However, there are infinite different motions and disturbances that can show up in real-world scenarios so that it is practically infeasible to record a realistic training data set.

The approaches presented in this section try to combine deep metric learning and the end-to-end classifiers to have on the one hand the advantages of metric learning and on the other hand to directly get the class scores from the network without any postprocessing. There exist several end-to-end learning approaches that follow the aforementioned idea. Most of the ideas were developed for face verification task. For face verification, the objective is very similar to an open-set classification. Imagine a verification system at the entrance of an office. Then the employees are the set of known classes and an intruder is a sample of an unknown class, which has to be identified as such. As you notice, it is infeasible to train the network with faces of all existing people; therefore, one promising option is to use metric learning. In the following text, several end-to-end metric learning approaches based on the cosine similarity followed by several approaches based on the Euclidean distance are presented.

2.3.1 Cosine Similarity

The cosine similarity is the natural metric in a conventional classification network where the last layer is a fully connected layer with softmax activation. The softmax activation is a normalized exponential function that normalizes its input to a probability distribution. The output of the fully connected layer is defined as follows:

$$\mathbf{x}_{\text{out}} = \mathbf{W}\mathbf{x}_{\text{in}} + \mathbf{b} \tag{2.8}$$

where $\mathbf{x}_{\text{in}} \in$ is the input vector, b is the bias vector, and W is the weight matrix. The score of class i is defined as follows:

$$\mathbf{x}_i = \mathbf{w}_i\mathbf{x}_{\text{in}} + b_i$$
$$\mathbf{x}_i = \|\mathbf{w}_i\|\|\mathbf{x}_{\text{in}}\| \cos(\theta_i) + b_i \tag{2.9}$$

where $\mathbf{w}_i$ is the ith column vector of the weight matrix W and the decision boundary between two classes is defined as follows:

$$\|\mathbf{w}_i\|\|\mathbf{x}_{\text{in}}\| \cos(\theta_i) + b_i = \|W_j\|\|\mathbf{x}_{\text{in}}\| \cos(\theta_j) + b_j \tag{2.10}$$

In [16], the normalization of the classifier layer was proposed. The biases are removed, and the weight vectors are L_2 normalized, so that the embedded features lie on a hypershpere. By doing this normalization, the class score is directly given as the cosine similarity between input vector x and the respective class weight vector. The decision boundary therefore reduces to

$$\|\mathbf{x}_{in}\| \cos(\theta_i) = \|\mathbf{x}_{in}\| \cos(\theta_j)$$
$$\cos(\theta_i) = \cos(\theta_j) \tag{2.11}$$

which is the cosine similarity between the embedded feature vector x and the weight vector of the respective class. In [16], the superior classification performance compared to the conventional softmax is demonstrated. Thus, all of the presented metric learning approaches in this section are based on the normalized softmax. It does not only improve the classification accuracy but additionally makes the classification more interpretable since the classification score is directly given by the angle. The softmax activation is given by

$$f_{\text{softmax}} = \frac{\exp(z_i)}{\sum_{j=0}^{|Y|} \exp(z_j)} \tag{2.12}$$

and the resulting default softmax cross entropy loss is given by

$$L_{\text{softmax}} = -\log\left(\frac{\exp(\cos(\theta_i))}{\sum_{j=0}^{|Y|} \exp(\cos(\theta_j))}\right) \tag{2.13}$$

where z_i is the class score, which in this case is the cosine similarity.

2.3.1.1 Multiplicative Margin – SphereFace

The first paper that aims to add a margin to the softmax classification in order to make the class features more discriminative was published in 2017 [17]. A margin was introduced in a multiplicative way. The angle to the true class has to be m times smaller than to all the other negative classes. The resulting loss is therefore given as follows:

$$L_{\text{sphere}} = -\log\left(\frac{\exp(\cos(m\theta_i))}{\exp(\cos(m\theta_i)) + \sum_{j=0,j\neq i}^{|Y|} \exp(\cos(\theta_j))}\right) \tag{2.14}$$

where i indicates the true class label. The loss defined in Eq. (2.14) is only valid for $\theta_i \in [0, \frac{2\pi}{m}]$ due to the periodical behavior of the cosine function. The authors of SphereFace generalize the loss to

$$L_{\text{sphere}} = -\log\left(\frac{\exp(\phi(\theta_i))}{\exp(\phi(\theta_i)) + \sum_{j=0,j\neq i}^{|Y|} \exp(\cos(\theta_j))}\right) \tag{2.15}$$

where $\phi(\theta_i) = (-1)^k \cos(m\theta_i) - 2k$, $\theta_i \in \left[\frac{k\pi}{m}, \frac{(k+1)\pi}{m}\right]$ and $k \in [0, m-1]$. $\phi(\theta)$ is a monotonically decreasing function.

2.3.1.2 Additive Margin – CosFace

The approach called CosFace follows a similar idea. Instead of adding a margin as a multiplicative factor, the authors of [18] propose to add a margin directly onto the cosine similarity between the embedded feature vector of the input and the corresponding weight vector. The resulting loss can therefore be written as follows:

$$L_{\cos} = -\log\left(\frac{\exp(\cos(\theta_i) + m)}{\exp(\cos(\theta_i + m)) + \sum_{j=0, j\neq i}^{|Y|} \exp(\cos(\theta_j))}\right) \tag{2.16}$$

2.3.1.3 ArcFace

A third approach adding a margin to normalized softmax classification was published in 2019 and is called ArcFace [19]. The idea is to add a margin directly to the angle instead of the cosine similarity between the input feature vector and the respective class weight vector. Therefore, the additive margin is within the cosine term and thus the loss is given by

$$L_{\text{arc}} = -\log\left(\frac{\exp(\cos(\theta_i + m))}{\exp(\cos(\theta_i + m)) + \sum_{j=0, j\neq i}^{|Y|} \exp(\cos(\theta_j))}\right) \tag{2.17}$$

Further in the ArcFace paper, a combination of the margin terms proposed in SphereFace, CosFace, and ArcFace can be used. The loss function therefore is given as follows:

$$L_{\text{arc}} = -\log\left(\frac{\exp(m_1 \cos(\theta_i + m_2)) + m_3}{\exp(m_1 \cos(\theta_i + m_2)) + m_3 + \sum_{j=0, j\neq i}^{|Y|} \exp(\cos(\theta_j))}\right) \tag{2.18}$$

It was shown in the paper that this multimargin loss term can also lead to very performative results.

2.3.1.4 Summary of Faces

The SphereFace, CosFace, and ArcFace approaches were published shortly after each other and are very related. All losses were proposed in the field of face verification, and all of them are basing on the normalized softmax and try to introduce a margin between the classes. For all approaches, the objective of adding a margin is to separate the classes well in order to be able to detect an outlier or more application-specific verify if a face is one of the known ones or not. The first row of images in Figure 2.3 provides an overview of the different angle or cosine similarity-based approaches introducing a margin. The second row shows the corresponding margins when using an Euclidean distance. These methods are discussed in the next section. Please note that in the first row, the *x*- and *y*-axes are the angles between an input sample and the class weight vectors of class one and two, whereas in the second row, the coordinate axes are the Euclidean distances.

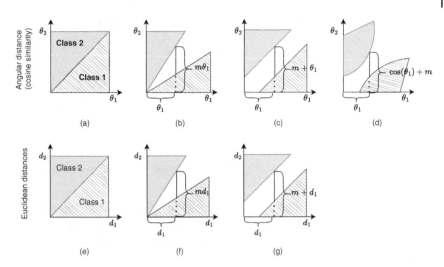

Figure 2.3 Different types of margins with regards to angular distances and Euclidean distances. The filled areas indicate where a test sample is assigned to class 1 whereas the shaded areas indicate where a test sample is assigned to class 2. If a test sample has certain distances to class 1 and class 2, so that it is in the white area, it is not assigned to any class, but detected as an outlier. (a) Softmax, (b) SphereFace, (c) ArcFace, (d) CosFace, (e) nearest cluster, (f) multiplicative margin, and (g) additive margin.

Figure 2.3a shows the separation between two classes in the normalized softmax case. Both classes are directly next to each other. If θ_1 is larger than θ_2, then the sample is assigned to class 1 and the other way round. There is no undefined space between these classes, which would make it possible to decide that a sample does not belong to either class. In Figure 2.3b, the multiplicative margin of SphereFace is illustrated. The larger the angle gets to the corresponding class, the larger the required margin gets between the classes. The ArcFace loss is depicted in Figure 2.3c. It is an additive margin to the angle and thus has a constant width. Figure 2.3d shows the CosFace loss. Please note, that unlike in the previous subplots, the x- and y-axis is not θ_1 and θ_2, but $\cos(\theta_1)$ and $\cos(\theta_2)$, which leads to a nonlinear dependency on the angular distance. Furthermore, as mentioned in the previous subsection, any combinations of the margin are also possible.

2.3.1.5 D-Softmax

The D-Softmax [20] differs from the previous approaches in that it does not use any margin. Instead, the authors of the D-Softmax loss disentangle the softmax loss function into an explicit intraclass loss part and interclass loss part. Both loss terms are minimized during training which means that the spread of the embedded feature vectors of the same class is minimized, whereas the distance between the classes is maximized. While the margin-based loss functions are satisfied if the

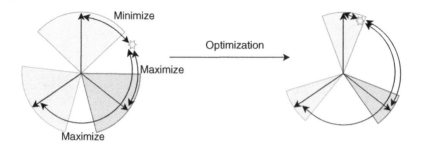

Figure 2.4 Visualization of the explicit intra- and interclass optimization of softmax dissection.

classes are more far away then a certain margin, the D-Softmax continuously tries to further separate the classes and to reduce the intraclass spread as visualized in Figure 2.4. This in the first place is a good characteristic; however, in practice, it is prone to overfitting.

The loss is composed of two components: First, there is the intraclass part z_i that denotes the score of the true class of the training sample; and second, there is the interclass part given as the sum of all other class scores $\sum_{j=0,j\neq i}^{|Y|} \exp(\cos(\theta_j))$. The authors of [20] propose to set the interclass loss part to a fixed value ϵ. Therefore, the intraclass loss part can be described as follows:

$$
\begin{aligned}
L_{d,\,intra} &= -\log\left(\frac{\exp(\cos(\theta_i))}{\exp(\cos(\theta_i)) + \epsilon}\right) \\
&= \log\left(\frac{\exp(\cos(\theta_i)) + \epsilon}{\exp(\cos(\theta_i))}\right) \\
&= \log\left(1 + \frac{\epsilon}{\exp(\cos(\theta_i))}\right)
\end{aligned}
\tag{2.19}
$$

and on the other hand, the positive class score is set to one for the explicit interclass loss, that is given as follows:

$$
\begin{aligned}
L_{d,\,inter} &= -\log\left(\frac{1}{1 + \sum_{j=0,j\neq i}^{|Y|} \exp(\cos(\theta_j))}\right) \\
&= \log\left(1 + \sum_{j=0,j\neq i}^{|Y|} \exp(\cos(\theta_j))\right)
\end{aligned}
\tag{2.20}
$$

The overall D-Softmax loss is therefore given as follows:

$$
\begin{aligned}
L_d = L_{d,\,intra} + L_{d,\,inter} &= \log\left(1 + \frac{\epsilon}{\exp(\cos(\theta_i))}\right) \\
&+ \log\left(1 + \sum_{j=0,j\neq i}^{|Y|} \exp(\cos(\theta_j))\right)
\end{aligned}
\tag{2.21}
$$

2.3.1.6 Softmax Center-Loss

Similar to the D-Softmax, the softmax center-loss [21] tries to optimize the interclass separability and intraclass compactness explicitly. It is composed of a softmax and cross-entropy loss term that causes the classes to be separated in the embedded feature space and the center-loss that aims to minimize the Euclidean spread within the classes. The loss function of the softmax center loss is given as follows:

$$L_{s,\,center} = -\log\left(\frac{\exp(\cos(\theta_i))}{\sum_{j=0}^{|Y|} \exp(\cos(\theta_j))}\right) + \lambda\|x_i - c_i\|_2^2 \tag{2.22}$$

where c_i is the mean feature vector of the corresponding class. The mean feature vectors of all classes are recalculated after every epoch using the embedded feature vectors of all training samples of the corresponding class. The effect of the center-loss training depending on the temperature factor λ is shown in Figure 2.5. The data set used in this experiment is the Modified National Institute of Standards and Technology (MNIST) data set. Using a deep convolutional neural network (CNN), the images were projected to a two-dimensional feature space and visualized after training the network with center loss using different values of λ which is basically controlling the impact of the center-loss term. The circular distribution of the class clusters is caused by the final fully connected layer, which outputs cosine similarity scores, in addition with softmax. However, when increasing the impact of the center-loss term, the clusters become more compact with respect to the Euclidean distance to their class clusters. Further, it is worth noting that the spread in radial direction is much more reduced than in the angular direction.

Both loss parts act on different metrics. The class scores are based on the cosine similarity whereas the center loss optimizes the Euclidean spread of the clusters. This does not necessarily mean that the class clusters are more compact in the angular way that finally does the classification.

2.3.2 Euclidean Distance

The earlier discussed end-to-end metric learning approaches are based on the cosine similarity since it is the natural score when doing the vector matrix multiplication in a fully connected layer. However, in theory, every other metric could also be used. The Euclidean distance is for example a very prominent one. In low-dimensional spaces, it is our intuitive distance metric; however, in high-dimensional spaces, it suffers from the curse of dimensionality. Loosely speaking, the curse of dimensionality says that for different kind of randomly distributed data sets, the maximum distance and the minimum distance

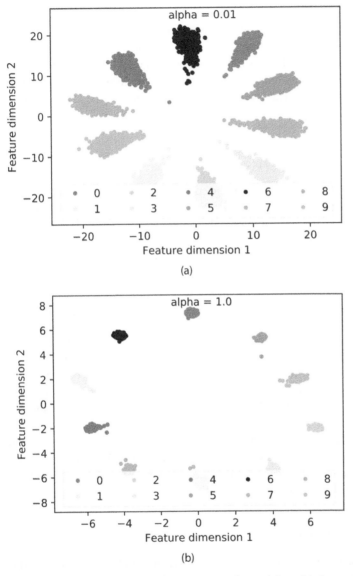

Figure 2.5 Visualization of the feature space after training with the center with (a) $\lambda = 0.01$ and (b) $\lambda = 1.0$.

within points in the data set get closer, the higher the dimensionality gets. In mathematically terms, it can be written as follows:

$$\lim_{d \to \inf} = \frac{dist_{max} - dist_{min}}{dist_{min}} \to 0 \tag{2.23}$$

where d is the dimensionality. This fact makes it difficult for algorithms that heavily base on the Euclidean distances between vectors to work reliably in high-dimensional spaces. First of all this is obviously a bad prerequisite for using the Euclidean distance for metric learning. However, in recent research, it was shown that this is only the case for uncorrelated data, when attributes are correlated on the other side it even gets easier to separate data points [22, 23]. Dimensions that contain information will improve the clustering, whereas irrelevant dimensions will decrease the performance. Thus, as long as the signal-to-noise ratio of the feature vectors, where signal is represented by the dimensions holding information and noise refers to dimensions that do not contain relevant information for classification, the Euclidean distance works reliable for clustering tasks. From a practical perspective, methods like the triplet loss also work on the Euclidean distance and achieved very good results in high-dimensional spaces. Additionally, for a short-range radar application, we are aiming for small and efficient network architectures since there are a lot of applications in resource-constraint environments such as in wearable devices, in light switches, or as a small device attached to the ceiling for people counting. Therefore, the curse of dimensionality gets an even minor problem.

In order to enable an end-to-end Euclidean distance-based architecture, the conventionally used fully connected layer as the last layer has to be adapted. In [24], an Euclidean distance layer is proposed. Instead of doing a matrix multiplication, the input feature vector is columnwise subtracted from the weight matrix and subsequently, the Euclidean norm of each difference vector is provided as the output. Therefore, the output of the Euclidean distance layer is given as follows:

$$p_i = \|\mathbf{x} - \mathbf{w_i}\|_2 \tag{2.24}$$

where i is the class index. The weight vector $\mathbf{w_i}$ can be interpreted as the center vector of class i. During training, the weights are optimized, and thus, the class centers shift to optimal positions in the Euclidean space.

2.3.2.1 Direct Optimization

For end-to-end learning approaches, usually the softmax is applied that turns the classification in a closed-set classification as previously discussed. However, the classifier can also be optimized directly on the Euclidean distances without mapping the scores to a probability distribution between the known classes. The objective is then to minimize the Euclidean distance to the positive class center

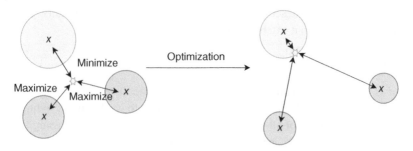

Figure 2.6 Visualization of the direct optimization of Euclidean distances to the different class cluster centers. Distance to the label class is minimized, whereas the distance to all other classes is maximized.

and to maximize the Euclidean distance to all negative class centers. The working principle is visualized in Figure 2.6. A possible loss function can be describe as follows:

$$L_{\text{eucl}} = L_{\text{eucl, intra}} + L_{\text{eucl, inter}} = \|x - w_1\|_2 + \sum_{j=0}^{|Y|} 1/\|x - w_2\|_2 \tag{2.25}$$

The presented loss function tries to continuously reduce the intraclass spread, which can easily lead to overfitting. Another approach, similar to triplet learning and the SphereFace, ArcFace, and CosFace, is to introduce a margin. When the margin condition is fulfilled, then the loss is zero. Therefore, the optimization stops when the clusters are sufficiently separated and does not force the clusters to be inappropriate small.

There are two possible ways to add a margin. First, the margin can be added to the Euclidean distance to the positive class center. The loss is minimized if the Euclidean distance plus a margin m is smaller than the distance to the closest negative class center. The loss can therefore be formulated as follows:

$$L_{\text{eucl, add}} = \max(0, \|x - w_{\text{p}}\|_2 + m - \|x - w_{\text{n, min}}\|_2) \tag{2.26}$$

where $w_{\text{n, min}}$ is the closest negative class center. Another possibility is to use a multiplicative margin, then the loss is minimized if the closest negative class center is further away than m times the positive class center. The resulting loss is given as follows:

$$L_{\text{eucl, mul}} = \max(0, m\|x - w_{\text{p}}\|_2 - \|x - w_{\text{n, min}}\|_2) \tag{2.27}$$

where $w_{\text{n, min}}$ is the closest negative class center. Alternatively, both margins can be used in combination as shown in Eq. (2.28). The additive and multiplicative Euclidean margins are depicted in Figure 2.3b and c below their respective angular margin counterparts. In Figure 2.3a, the decision boundary for assigning a test sample simply to the closest class cluster center is shown.

$$L_{\text{eucl, com}} = \max(0, m_1 \|x - w_1\|_2 + m_2 - \|x - w_{n,\text{min}}\|_2) \tag{2.28}$$

2.3.2.2 Euclidean Softmax

The cosine similarity describes a similarity between two vectors, and therefore, is high if two vectors are similar, and low, if they are dissimilar. The Euclidean distance on the contrary works the other way round. Low distances mean a high-classification confidence. Thus, the scores have to be inverted first to use the same softmax and cross-entropy learning as with cosine similarity. To achieve this, the authors of [24] propose to map the Euclidean distance to a similarity measure similar as the cosine similarity. Unlike for the cosine similarity, where the angle has a value range from 0 to 2π and is therefore nicely mapped by the cosine to values between 1 and -1, the value range of the Euclidean distance reaches from 0 to infinity. Therefore, the Euclidean distance has to be mapped to a value range of 0 to 2π first. The mapping proposed in [24] is done as follows:

$$f_{\text{eucl}}(d_{\text{eucl}}) = \frac{\pi}{1 + d_{\text{eucl}}} \tag{2.29}$$

The result of function $f_{\text{eucl}}(d_{\text{eucl}})$ is already a similarity score. High distances are mapped to small values, whereas small distances are mapped to high scores. However, to make the Euclidean distance-based similarity score as similar as the cosine similarity, a cosine is applied to map the scores also to a value range from -1 to 1, where 1 is the highest similarity and -1 the lowest similarity. This is done as described in Eq. (2.30).

$$z_{\text{eucl}}(d_{\text{eucl}}) = \cos(f_{\text{eucl}}(d_{\text{eucl}}) + \pi) \tag{2.30}$$

The Euclidean softmax loss is then defined as follows:

$$L_{\text{eucl soft}} = -\log\left(\frac{\exp(z_{\text{eucl},i})}{\sum_{j=0}^{|Y|} \exp(z_{\text{eucl},i})}\right) \tag{2.31}$$

Further, in [24], it is proposed to use the Euclidean softmax in combination with the center-loss. The final loss function is given as follows:

$$L_{\text{eucl soft, center}} = -\log\left(\frac{\exp(z_{\text{eucl},i})}{\sum_{j=0}^{|Y|} \exp(z_{\text{eucl},i})}\right) + \lambda\|\mathbf{x} - \mathbf{w_i}\|_2^2 \tag{2.32}$$

where λ is a temperature value controlling the impact of the center-loss part. This loss is quite similar as the softmax center loss from Section 2.3.1.6 except of the important difference, that both loss parts, softmax, and center loss are now based on the same metric, namely the Euclidean distance.

Another variant of the Euclidean softmax is to add margins similar to the Sphere-, Arc-, and CosFace approaches. The Euclidean softmax loss including the margins can be written as follows:

$$
L_{\text{eucl soft, margin}} =
$$
$$
- \log \left(\frac{\exp(m_1 \cos(\|\mathbf{x} - \mathbf{w_i}\|_2 + m_2)) + m_3}{\exp(m_1 \cos(\|\mathbf{x} - \mathbf{w_i}\|_2 + m_2)) + m_3 + \sum_{j=0,j\neq i}^{|Y|} \exp(\cos(\|\mathbf{x} - \mathbf{w_j}\|_2))} \right)
$$
$$
(2.33)
$$

2.3.3 Summary

In Section 2.3.1, metric learning using a normalized fully connected layer and therefore, using the cosine similarity is discussed, whereas in Section 2.3.2, a Euclidean distance layer is used as last layer which results in Euclidean distance-based class scores. It is important to understand that the cosine similarity is a similarity score, whereas the Euclidean distance is a distance metric. Similarity and distance behave inversely to each other. If the similarity of a test sample to a certain class is small, then the test sample belongs with a high probability to this class, whereas when the Euclidean distance is large, it belongs with a small probability to this class. The ideal output using the cosine similarity is therefore a one-hot encoded vector with maximum value for the predicted class and minimum values for all other classes. Thus, conventionally a softmax normalization in addition to evaluating the cross-entropy between the softmax normalized output vector and the one-hot encoded label vector is also used as a loss function.

Contrary to this, the classifiers using an Euclidean distance layer outputs a distance which works inversely to the cosine similarity score. Thus, softmax normalization and cross-entropy are not directly applicable. Therefore, the natural approach is to optimize the Euclidean distances directly as it is presented in Section 2.3.2.1. Optimizing the Euclidean distances directly was already successfully used for pairwise methods. However, it is worth mentioning that in both cases, the respective conversion from distance metric to similarity score and the other way round can be done.

Cosine Similarity to Distance: For the cosine similarity, the conversion to distance is rather simple. It was actually already done, albeit indirectly, in the ArcFace work. The margin is not applied on the cosine similarity, but on the angle, which is a distance. The conversion to angle is simply done by applying the inverse function of cos(), namely arccos(), to the similarity score. Thus, the angle is given as $\theta = \arccos(\cos(\theta))$.

Euclidean Distance to Similarity: The conversion from Euclidean distance to a similarity score is a little bit more challenging. There are multiple ways to do this conversion. One way of doing the conversion to a similar similarity score as the cosine similarity was already presented in Section 2.3.2.2.

The output of both the layers, a conventional fully connected layer and an Euclidean distance classification layer, can be interpreted as a similarity score or a distance score. For using a similarity score, the conventional way is to apply softmax normalization along with cross-entropy loss. When using a distance metric, the conventional approach is to directly optimize the distance.

The softmax activation (2.33) normalizes the class scores to a probability distribution. The sum of all class probabilities sum up to one. Even though a test sample has a much lower similarity to every class as the average absolute similarity of samples within this class – and therefore is an outlier – after softmax normalization, it might happen that the class probability of one class is very high, when the similarity to the other classes is even lower. This issues is visualized in Figure 2.7. Although the distance to class 1 is in both cases the same, the scores after softmax normalization highly differ due to the larger distances of classes 2 and 3. The relative class score makes perfect sense for a closed-set classification; however, the question arises if a softmax normalization is the right tool for an open-set classification task or if it would be better to optimize the absolute distances, that are not normalized, directly. Unfortunately, no such experiments have been presented in the literature so far. Most metric learning approaches using a fully connected layer are using the similarity score and normalized softmax.

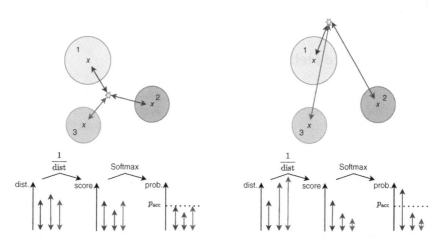

Figure 2.7 Visualization of the normalization issue when considering an open-set classification problem in the Euclidean space. Although the distance to class 1 is the same in both the scenarios, the relative class scores after softmax might highly vary due to the much larger distances to classes 2 and 3 in the right sketch.

For the Euclidean distance, there are methods to optimize the distance directly and convert it to a similarity score and apply it to softmax activation; however, also here no direct comparison in terms of open-set classification was done until now.

2.4 Proxy Methods

The proxy methods [25–27] are rather new deep metric learning approaches and originate from the pairwise metric learning branch. The idea is to learn similarities or distances to representative proxy vectors of the classes instead of learning similarities of distances between the single samples. The motivation is to mitigate the issues of the pairwise methods such as the exponential combinatorial possibilities to form pairs, triplets, quadruplets, and the mining of hard negatives. In general, it can be seen that over time, the pairwise methods try to incorporate more and more samples, starting from the contrastive loss over triplet loss, quadruplet loss up to *N*-pair loss. The logical consequences are the proxy methods where the proxies are representative feature vectors of an entire class. Therefore, in a wide sense, all samples of the respective class are regarded. For supervised classification, a proxy vector has exactly the same meaning as the class weight vectors in the last layer of the end-to-end learning approaches. This trend over multiple years of research let us strongly assume the superiority of end-to-end metric learning methods for supervised classification tasks. This is also underlined by the experiments in [19], where a direct comparison between Sphere-, Cos-, ArcFace, and triplet loss was done.

2.5 Advanced Methods

So far the basic deep metric learning concepts were introduced and discussed. Of course, there exist many variants building on top of these concepts. Presenting all of them would be beyond the scope of this book. However, we would like to give a glimpse of advances in deep metric learning by presenting selected approaches.

2.5.1 Statistical Distance

In the end-to-end learning approaches, the distances to the class weight vectors or class centers are used. However, the class cluster is not a single point but actually a distribution of feature vectors. From this fact, there arise two issues. The first one regards the intraclass distance. The metric learning approaches do not guarantee that the class clusters have equal dispersion in all feature directions. The clusters are typically not ideally hyperspherical which is actually assumed for the Euclidean distance. The second issue relates to the interclass distances. Typically, a test sample is detected as outlier when it is too far away to each of the classes. However, the data of some classes might have a higher variance than other class clusters, and thus it is hard to choose a good distance threshold for rejection for all classes.

To overcome these issues, some approaches propose to evaluate the class score by using a statistical distance between test sample and class clusters. The easiest approach is to use the Mahalanobis distance as defined in Eq. (2.1) and define M as the inverse class covariance matrix for each class. A covariance matrix for each class is estimated, and thus, there exists an individual distance metric. The usage of the described Mahalanobis distances solves the two discussed drawbacks, because the scaling with the inverse covariance matrices normalize the class clusters. It compensates the different dispersion in the different dimensions of each class clusters and additionally normalizes the variance of the different class clusters. Thus, a common rejection distance threshold can be defined without making compromises for any class.

2.5.1.1 Gaussian Classifier

In [28], a large margin Gaussian mixture loss is proposed. The fundamental assumption of this work is that the class clusters are Gaussian distributed. A multivariate Gaussian distribution is defined by the covariance matrix and the mean vector. Both are estimated after each epoch, and the class scores are then defined as follows:

$$
\begin{aligned}
f_c(x) &= \frac{1}{\sqrt{(2\pi)^d det(\Sigma_c)}} e^{-\frac{1}{2}(x-\mu)^T \Sigma_c^{-1}(x-\mu)} \\
&= p_c e^{-\frac{1}{2}d_c}
\end{aligned}
\tag{2.34}
$$

where p_c is a class-specific prefactor and d_c is the Mahalanobis distance from input sample x to the class distribution of class c. To better separate the classes, a margin is added to the Mahalanobis distance. Similar to the softmax, the authors propose to normalize the output scores to a probability distribution. The classification loss is similar to the softmax loss and given as follows:

$$
L_{cls} = -\log \frac{p_i e^{-\frac{1}{2}(d_i+m)}}{p_i e^{-\frac{1}{2}(d_i+m)} + \sum_{j=0, j\neq i}^{|Y|} p_j e^{-\frac{1}{2}d_j}}
\tag{2.35}
$$

$$
L_{lkd} = -\log \mathcal{N}(x, \mu_i, \Sigma_i)
\tag{2.36}
$$

Further a second loss term, Eq. (2.36) is used that optimizes the log-likelihood of input sample x to its assigned class distribution. It is pointed out in the research that this is closely related to the center loss formulation and thus optimizing the intraclass distance. The final large Gaussian mixture loss is a combination of L_{cls}

and L_{lkd}, where the impact of L_{lkd} is weighted by a temperature factor λ. The total loss is thus given as follows:

$$L_{LGM} = L_{cls} + \lambda L_{lkd} \tag{2.37}$$

For the Mahalanobis distance, the covariance matrix has to be inverted. This is only possible if the covariance matrix is nonsingular, which is not always valid. Thus, the authors of [28] approximate the covariance matrix as diagonal matrix, and then the inversion is simply the inverted diagonal elements. However, this is a quite strong assumption especially when the training does not aim for a diagonalization of the covariance matrices.

2.5.1.2 Statistical Triplets

In the large margin Gaussian mixture loss, the covariance matrices are explicitly estimated after each epoch. Another way to approximate the shape of the clusters is to use a variational auto-encoder (VAE) in combination with metric learning. A simple auto-encoder architecture projects the input to an embedding space and thus encodes the data as a feature vector in the first part of the network, which is also called the encoder. In the second part of the network, which is basically the decoder, the original input is tried to be reconstructed. Therefore, the network learns to compress the input data in a way that the reconstruction error is minimal and thus the most important information of the input is preserved. A drawback of this approach is that the latent space might be badly conditioned. The data set consists of a finite set of samples that are projected to a finite set of embedding vectors. Therefore, it is only ensured that outgoing from these embedding vectors, the decoder can reconstruct something reasonable. However, it might happen that the network drastically overfits and an embedded feature vector in the close neighborhood is reconstructed to something completely different. In other words, since the latent space is defined by a finite set of discrete vectors, no continuity is ensured.

A VAE is basically doing the same as an auto-encoder, except the fact that the input is not compressed to a single feature vector, but to a feature distribution. The vector that is used for reconstruction is sampled from this feature distribution. Therefore, each time a slightly different feature vector is used for reconstruction. Since the reconstruction label stays the same, it is ensured that all feature vectors in the near neighborhood are reconstructed to the same data. This makes the embedding space smooth and continuous. Since also for metric learning, the goal is to learn a continuous metric space, and the VAE architecture is well suited. In addition to the fact that the embedding space gets more continuous, there comes an additional favorable property with the VAE. The embedded feature distributions can be seen as independent, since for example the execution of one gesture does not depend on how the previous gesture was performed and identically distributed. Due to the independent and identically distributed (iid) assumption, the class distribution can easily be estimated as the mean of the means and the mean of the variances of all sample feature distributions of the class of interest.

The application at the end of this chapter is based on this method. Therefore, further details about implementation and loss function are discussed in the application section.

2.5.2 Structured Metric Learning

Label-Aware Ranked Loss: The label-aware ranked (LAR) loss is a novel metric loss function. The state-of-the-art deep metric learning losses we discussed so far have only two possible similarity scores. If either of the two samples are from the same class, then they are similar, or from different classes, then they are equally dissimilar. However, there are classification tasks where the classes are not equally dissimilar. A very good example is radar-based people counting. It is a classification task where the predicted class equals the number of people in the scene. Obviously mis-classifying by one person is better than mis-classifying by two or more persons. Therefore, the LAR loss takes advantage of the ranked ordering of the labels in regression problems. In [29], it is shown that the loss minimizes when data points of different labels are ranked and laid at uniform angles between each other in the embedding space. The loss function is defined as follows:

$$L_{\text{LAR}} = \frac{1}{N} \sum_{i=1}^{N} \log \left(1 + \sum_{j \neq i} \exp \left(\log(\Delta_l) f_i^{aT}, f_j^n - f_i^{aT} f_j^p \right) \right) \tag{2.38}$$

where

$$\Delta_1 = \min \left(|t_a - t_n|, |L - |t_a - t_n|| \right) \tag{2.39}$$

The loss uses the multiplier $\log(\Delta_l)$ to regulate the ranking of the labels. Here, t_a is the label of the anchor, t_n is the label of the current negative sample, and L is the number of different labels.

The minimizer of the LAR loss for six labels is shown in Figure 2.8 with the respective label in the circle as well as the Δ_l multiplier assignments, with respect to the first label, on the edges. Experimental results show that LAR loss is superior to other metric learning losses, when the data have an implicit ranking, e.g., counting people.

Hierarchical Approaches: Another group of approaches that try to improve the metric learning by structuring the embedding space are the hierarchical methods. Especially for classification tasks with many classes, there are a huge number of very distinct classes, and thus there exist many easy triplets where the triplet margin condition is fulfilled leading to a vanishing loss. As a result, the learning speed drastically decays. Therefore, the authors of the hierarchical

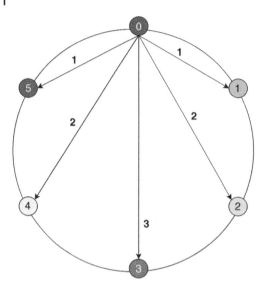

Figure 2.8 Optimal 2D LAR loss label positions (circles) and multiplier assignments (edges) for six labels.

triplet loss [30] propose to structure the classes in a hierarchical way and adapt the margin value depending on which level two classes are related. In [31], the usage of hierarchical proxies is proposed. On the lowest level, a proxy represents a single class. Higher-level proxies represent a number of lower-level proxies. A training sample aims to reduce the distance on each level to the proxy it is assigned to and to increase the distance on each level to all proxies it is not assigned to. By this the authors intend to better represent semantic similarities in the learned distances.

2.6 Application: Gesture Sensing

Gesture sensing is one of the applications that profit a lot from metric learning. It is desired to have a high-recognition rate on the valid and predefined gestures, while being robust against alien motions. In other words, motions that are not intended to be a gesture should be detected as such and not misclassified as one of the valid gestures. The gestures that should be recognized by the system can be predefined and multiple repetitions from multiple user can be recorded. However, the alien motions can literally be anything. Therefore, it is almost infeasible to record a good representative data set for alien motions. Metric learning helps to identify alien motions even by training the network solely with samples from valid gestures.

The presented application was also published in [32] and follows the idea outlined in Section 2.5.1.2 and, therefore, uses a VAE architecture that is trained with a novel metric learning technique. The section is structured as follows: first, the radar system and its configuration is introduced. Second, the data set including the preprocessing of the radar data, and the used reconstruction labels is discussed.

Third, the architecture is presented in more detail, and the loss function and the training procedure are described. As a last point, the training results are presented.

2.6.1 Radar System Design

The 60-GHz BGT60TR13C frequency modulated continuous wave (FMCW) radar chipset from Infineon is used for the gesture recognition system presented in this section. The radar is configured with the values given in Table 2.1. With this radar configuration, the radar is able to resolve targets up to a range of 0.96 m with a resolution of 3.0 cm, and from the chirp timing results, a maximum resolvable radial velocity range of 3.205 m/s. This includes positive as well as negative velocities. Using 32 chirps per frame, the velocity can be resolved by a fast-Fourier transform (FFT) up to 0.1 m/s.

2.6.2 Data Set and Preparation

Before the training of the network can be started, a data set is needed. Further, for deep learning based on radar data, the data is conventionally preprocessed. Therefore, in this chapter, we will first introduce the set of gestures that our system should be able to recognize and then explain the preprocessing of the radar data and the generation of the reconstruction labels.

2.6.2.1 Gesture Set

The set of gestures includes the 10 different hand gestures depicted in Figure 2.9. The radar is placed on the table with its antennas facing upward. The gestures are

Table 2.1 Operating parameters of the used radar chipset BGT60TR13C.

Parameters	Symbol	Value
Ramp start frequency	f_{min}	58 GHz
Ramp stop frequency	f_{max}	63 GHz
Bandwidth	B	5 GHz
Range resolution	δr	3.0 cm
Number of samples per chirp	N_s	64
Maximum range	R_{max}	0.96 m
Chirp repetition time	T_{PRT}	0.39 ms
Maximum Doppler	v_{max}	3.205 ms^{-1}
Number of chirps per frame	N_c	32
Doppler resolution	δv	0.2 ms^{-1}
Number of Tx antennas	N_{Tx}	1
Number of Rx antennas	N_{Rx}	3

PRT, pulse repetition time.

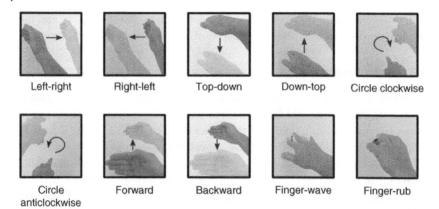

Left-right	Right-left	Top-down	Down-top	Circle clockwise

Circle anticlockwise	Forward	Backward	Finger-wave	Finger-rub

Figure 2.9 Overview of performed gestures.

performed in a short-range up to 30 cm above the radar in the total eight people volunteered to record data for training and testing purposes. For data recording, a small tool was used that randomly displays a gesture to be performed by a user. Once the user confirms its readiness, a green light shows up, and the next two seconds are recorded and saved. Within this time, the user has to perform the gesture in short-range above the radar. The recording was performed under supervision. The supervisor rejected the recording, if a wrong gesture was executed.

2.6.2.2 Data Preparation

Many gestures such as the swipes require much less time than two seconds. Therefore, after recording, a start and end of gesture is detected based on a simple energy thresholding on the raw data. Only the signal where a hand motion is detected has to be preprocessed. The remaining signal is set to zero.

The data preparation depends on the gesture set. The data set used in this work contains some gestures that are only distinguishable in range and Doppler such as the top-down and down-top gestures. Additionally, the set contains swipes in all possible directions. Since the swipes are symmetric motions, they look exactly the same in the range and Doppler spectrum no matter in which direction they are executed. Thus, to be able to distinguish all 10 gestures, an azimuth and an elevation estimation is additionally needed. Consequently, to uniquely distinguish the 10 different gestures, the range, Doppler, azimuth, and elevation information over time are required. Therefore, the data are preprocessed as range-, Doppler, azimuth-, and elevation-spectrograms to implement the presented gesture-sensing solution. The equations for generating the spectrograms from the raw data can be found in Chapter 1. An exemplary set of spectrograms is visualized for each macrogesture in Figure 2.10 and for each microgesture in Figure 2.11.

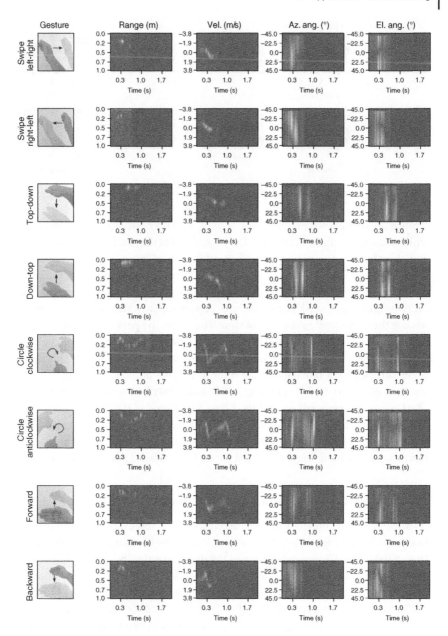

Figure 2.10 Exemplary set of spectrograms for each macrogesture in the data set.

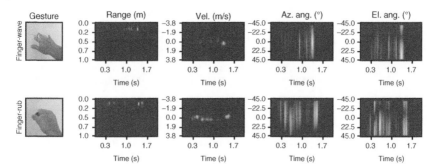

Figure 2.11 Exemplary set of spectrograms for each microgesture in the data set.

Since an auto-encoder architecture is used, we also need reconstruction labels, which are spectrograms in our case. Only the signal of the hand has to be preserved and therefore reconstructed. All other artifacts like noise or other reflections should actually even be removed. Therefore, the aim is to reconstruct the input spectrograms to filtered spectrograms where the hand signal is preserved, but everything else mitigated. For short-range gesture sensing, we can assume that the hand is the only target in the field of view. Therefore, the peak in each time step in the spectrograms is used as the signal reflected from the hand. Extracting the peak position in each timestep from the spectrograms results in a 1D signal. This signal is smoothened by a moving average filter to remove outliers. The spectrograms are then reconstructed by creating a Gaussian distribution in each time step. The Gaussian is centered around the filtered 1D signal of the peak positions and created so that the 95% confidence interval stretches over ±3 bins. The resulting filtered spectrograms or reconstruction labels are exemplarily shown for each macrohand gesture in Figure 2.12 and for each microhand gesture in Figure 2.13.

2.6.3 Architecture and Metric Learning Procedure

As discussed in the previous section each recording is represented by a set of spectrograms. A spectrogram is a 2D matrix or it can also be seen as grayscale image. Due to the image like representation of the data, we are using a 2D convolutional neural network architecture. The encoder-network consists of three convolutional layers using filter sizes of (5×5), (3×3), and (3×3) and 32, 32, and 64 channels followed by Dropout layers with rate 0.5. To reduce the data size, two max-pooling layers with pooling sizes (2,2) are added after the first two convolutional layers. Afterward the tensor is flattened and two parallel fully connected layers output a 32-dimensional mean vector and a 32-dimensional variance vector. The mean and the variance vector represent the embedded Gaussian distribution to which each input sample is projected. From this distribution, an embedded feature vector is

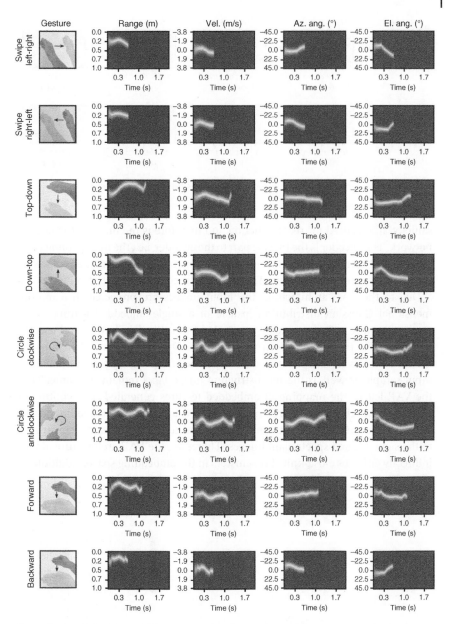

Figure 2.12 Exemplary set of filtered spectrograms for each macrogesture in the data set. A set of filtered spectrograms are used as reconstruction labels during training.

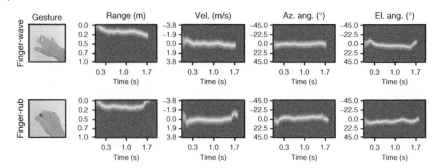

Figure 2.13 Exemplary set of filtered spectrograms for each microgesture in the data set.

sampled that is used for reconstruction and therefore is the input to the decoder network. The decoder is the inverse part to the encoder, i.e., the max-pooling layers are replaced by up-sampling layers, and the convolutional layers are replaced by transposed convolutional layers. The entire architecture is depicted in Figure 2.14.

Due to the VAE architecture, the input images are projected to a multidimensional Gaussian distribution instead of a single-embedded feature vector. During training, a feature vector is sampled from this distribution and used for reconstructing the filtered spectrogram images. Due to this generative behavior, the embedding vector used for reconstruction is different every time, although the input sample as well as the reconstruction image label remain the same. Thus, the VAE learns the mapping of embedded features generated by a continuous distribution to the same filtered image label. As a result, the embedding feature space is forced to be continuous and close-by embedded features are reconstructed to the same filtered spectrogam images. Therefore, the VAE architecture already indirectly enforces close-knit class-clusters in the embedding space. Additionally,

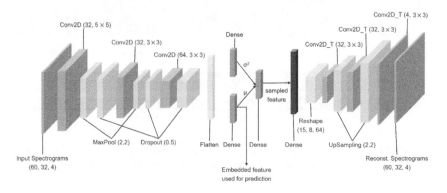

Figure 2.14 Architecture of the used VAE model.

due to triplet learning, the class clusters are pushed apart in the embedding space. Due to the generative aspect of the architecture, smooth and compact class clusters are obtained, whereas the triplet learning approach increases the distance between the class clusters. Therefore, a VAE architecture combined with triplet loss, which is referred to as triplet variational auto-encoder (TVAE) in this book, is well suited for detecting embedded features generated by background motion.

To train the TVAE, we propose the weighted sum of multiple objectives which are introduced in the following.

2.6.3.1 Statistical Distance Triplet Loss

The idea of the triplet loss is to feed three samples into the neural network. The first is the anchor, the second is a random sample of the same class, and the third is a random sample of any other class. The distance between anchor and either positive or negative sample is defined as follows:

$$d(x_1, x_2) = (x_1 - x_2)^\mathsf{T}(x_1 - x_2) \tag{2.40}$$

where x_1 is the anchor and x_2 is either the positive or negative sample. When using the VAE architecture, the embedding is modeled as Gaussian distribution. Thus, the Mahalanobis distance between the embedded anchor distribution and the mean of either the positive or negative sample and evaluated. Also, other statistical distance metrics between a point and a distribution or between two distributions like the Wasserstein metric can be used. The statistical distance based on the Mahalanobis distance is defined as follows:

$$d^{\mathrm{stat}}(\mu_a, \Sigma_a, \mu_2) = (\mu_a - \mu_2)^\mathsf{T}\Sigma_a^{-1}(\mu_a - \mu_2) \tag{2.41}$$

where μ_a and Σ_a are the mean and covariance matrix of the anchor distribution X_a and μ_2 is either the mean of the positive or negative sample distribution. The triplet and statistical distance triplet losses are finally defined as follows:

$$L_{\mathrm{triplet}} = \max\left(d(x_a, x_p) - d(x_a, x_n) + \alpha, 0\right) \tag{2.42}$$

or

$$L_{\mathrm{triplet}}^{\mathrm{stat}} = \max\left(d^{\mathrm{stat}}(\mu_a, \Sigma_a, \mu_p) - d^{\mathrm{stat}}(\mu_a, \Sigma_a, \mu_n) + \alpha, 0\right) \tag{2.43}$$

respectively, where μ_a and Σ_a define the anchor distribution X_a, μ_p, and μ_n are the mean feature vectors of positive and negative sample, respectively, and α is a hyperparameter. As a result, the triplet loss evaluates the distance between single embedded feature vectors of anchor, positive, and negative, whereas the statistical distance triplet loss evaluates the distance between the anchor distribution and the mean vector of positive and negative sample.

2.6.3.2 Reconstruction Loss

The reconstruction loss aims to minimize the difference between the reconstructed images and the label images. As metric, the mean squared error defined as follows:

$$L_{\text{MSE}} = \sum_{c=0}^{C-1} \sum_{n=0}^{N-1} \sum_{m=0}^{M-1} (Y_{\text{rec}} - Y_{\text{lab}})^2 \tag{2.44}$$

where C is the number of channels, N and M are the dimensions of the images, Y_{rec} are the reconstructed images, and Y_{lab} are the reconstruction labels, is used.

2.6.3.3 KL-Divergence Loss

For the VAE, the embedding of an input sample is modeled as a multivariate Gaussian distributed random variable X. The underlying and unknown distribution is approximated by a multivariate standard Gaussian distribution. The difference between the underlying distribution of the embedding and the multivariate standard Gaussian distribution is evaluated using the Kullback–Leibler divergence defined as follows:

$$L_{\text{KL}} = D_{\text{KL}}[N(\mu(X), \Sigma(X)) || N(0,1)] = \frac{1}{2} \sum_{k=0}^{K-1} (\Sigma(X)_k + \mu(X)_k^2 - 1 - \log \Sigma(X)_k)$$
$$\tag{2.45}$$

where K is the dimension of a random variable X and $\mu(X)_k$ and $\Sigma(X)_k$ is the mean and variance value of its k dimension. The resulting divergence defines the Kullback–Leibler (KL)-divergence loss. By optimizing the KL-divergence, the maximization of the variational lower bound is achieved.

2.6.3.4 Center Loss

The center loss minimizes the Euclidean intraclass distances, and therefore, leads to more discriminative classes. The center loss is defined as follows:

$$L_{\text{center}} = (\hat{\mu}_c - \mu_{c,i})^{\text{T}}(\hat{\mu}_c - \mu_{c,i}) \tag{2.46}$$

where $\hat{\mu}_c$ is the estimated mean of class c, and $\mu_{c,i}$ is the mean of the embedded feature distribution of a sample associated to class c. By minimizing the center loss, the intraclass spread is minimized and thus the class cluster gets more compact. If all samples of the same gesture are projected to a very compact area, then it gets easier to detect alien motions which are outliers.

2.6.3.5 TVAE Loss

The overall loss that is minimized during training the TVAE is defined as follows:

$$L_{\text{TVAE}} = \alpha_1 L_{\text{triplet}}^{\text{stat}} + \alpha_2 L_{\text{MSE}} + \alpha_3 L_{\text{KL}} + \alpha_4 L_{\text{center}} \tag{2.47}$$

where α_1 to α_4 are hyperparameters.

2.6.3.6 Class Scores

As already discussed in Section 2.5.1, the class clusters can have very different dispersion in different dimensions and, therefore, it is difficult to define a common rejection threshold for all classes. By estimating the covariance matrix of each class and using a class-dependent Mahalanobis distance, the distances and therefore class scores can be normalized. Under the assumption that each recording or sample is independent of each other and that the embedded distributions are identically distributed, the class distributions can be estimated by the mean of the embedded distribution of all samples associated with the same class. As a result, the mean of a class distribution is defined as follows: $\hat{\mu}_c = \frac{1}{|X_c|} \sum_{x \in X_c} \mu_x$ and the variance is defined as follows: $\hat{\sigma}_c^2 = \frac{1}{|X_c|^2} \sum_{x \in X_c} \sigma_x^2$ where X_c is the set of embedded feature distributions belonging to class c. The covariance matrix $\hat{\Sigma}_c$ is defined as a diagonal matrix with σ_c^2 as entries. Based on the estimated class distribution, the Mahalanobis distance can be used to evaluate the statistical distance defined as follows:

$$s_{c,i} = (\hat{\mu}_c - \mu_{c,i})^{\mathsf{T}} \hat{\Sigma}_c^{-1} (\hat{\mu}_c - \mu_{c,i}) \tag{2.48}$$

where $\hat{\mu}_c$ is the estimated mean of class c, $\hat{\Sigma}_c$ is the estimated covariance matrix of class c and μ_c is the embedded mean of a sample i belonging to class c. Due to the VAE architecture and the embedding to feature distributions it becomes especially simple to estimate the class distribution as shown above. This is another advantage of the VAE architecture.

2.6.4 Results

The proposed architecture is evaluated using the recorded test data set and compared to other state-of-art models. The state-of-art models we use for evaluation are the softmax classifier and a triplet-based model. Both models share the same architecture, which is the encoder of the TVAE. For the softmax model, a softmax layer was added which outputs the class scores. The class scores for the triplet model are determined by the Euclidean distance of the embedded feature vector of a test sample and the class cluster centers, which are estimated after each epoch using all training samples of the respective class.

For offline evaluation, a data set of 10 different gestures, as introduced in Section 2.6.2.1, from eight different individuals was recorded. In total, the data set contains 300 repetitions per gesture. Moreover, 575 samples of random movements or background noise for testing the robustness against false alarms were captured. The random movement data set contains on the one hand very different motions to the gestures but on the other hand also contains very similar samples like a diagonal hand movement or the circle gesture performed in the elevation angle direction instead of azimuth angle direction. Hence, those samples contain very similar features as the known gestures and are therefore especially hard to identify as alien motions.

2.6.4.1 Training

To preserve comparability of the different architecture, all networks are trained under the same conditions. All layers are initialized using tensorflow's default initializer "glorot uniform." Moreover, Adam optimizer with a learning rate of 0.001 is used. All networks, except of the softmax model which is trained using categorical cross entropy as defined in Eq. (2.13), are based on triplet loss and therefore, an anchor sample and randomly chosen positive and negative samples are fed into the network at a time. Each sample in a batch is once used as anchor. Due to the random selection of samples, the performance of the network may slightly differ from training to training. However, an epoch consists of 40 batches containing 60 samples. The networks are trained for 60 epochs. The data set was initially randomly shuffled and afterward split into five blocks, which are used to perform a fivefold cross validation. The samples within each block are the same in all training runs of the different networks.

The state-of-art triplet model is trained using the triplet loss as described in Section 2.2.2. The TVAE architecture is trained in a two-step procedure. For the first 50 epochs, the network is optimized using the loss L_{TVAE} with $\alpha_1 = \alpha_2 = \alpha_3 = 1$ and $\alpha_4 = 0$. Afterward for additional 10 epochs, α_1 to α_3 are set to 0 and α_4 is set to 1. Thus, the intraclass spread is additionally minimized to obtain even more compact class clusters.

2.6.4.2 Accuracy and F1-Score

The networks are trained with samples of only known gestures. However, to evaluate the rejection capability, unseen samples from known gestures as well as samples of nongestures were interpreted by the neural network. Afterward a threshold or distance in the embedding space determines whether a sample is accepted or rejected. The F1-score is used to evaluate the rejection capability, which is the harmonic mean of precision and recall. The rejection threshold is chosen as the one that maximizes the F1-score in the training set. In Table 2.2, the accuracy and F1-score of the different approaches is given. In column two and three, the accuracy and F1-score based on the test set including the background noise samples are shown. In the last column, the accuracy of the test set only including known gesture samples is given. The numbers show that all approaches perform very well in classifying explicitly known gestures. However, including the background or random samples, and therefore simulating a real-world situation, unveil the benefit of metric learning. Compared to the state-of-art softmax or triplet approaches, an increase up to about 13% in the F1-score was achieved.

Table 2.2 Accuracy and F1-scores.

Approach	Accuracy (%)	F1-score (%)	Accuracy (only known) (%)
Softmax	82.8	77.7	95.6
Triplet	82.9	76.8	95.0
TVAE center	90.3	90.8	97.6

2.6.4.3 Confusion Matrix

The confusion matrix (CM) shows the true label and the predicted class of all test samples. Therefore, it allows a detailed insight into the class-dependent classification performance. It can, e.g., unveil if there is a large confusion between some certain classes. In Figures 2.15 and 2.16, the confusion matrices of triplet and TVAE are shown. First of all, only very little confusion between the known

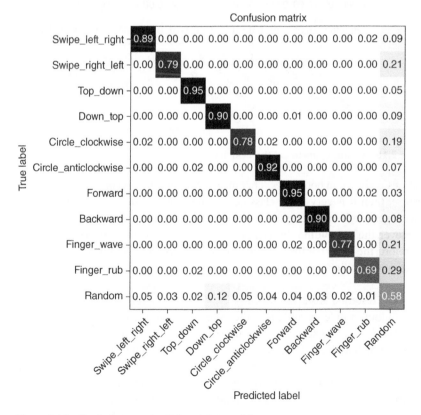

Figure 2.15 Confusion matrix of the triplet model.

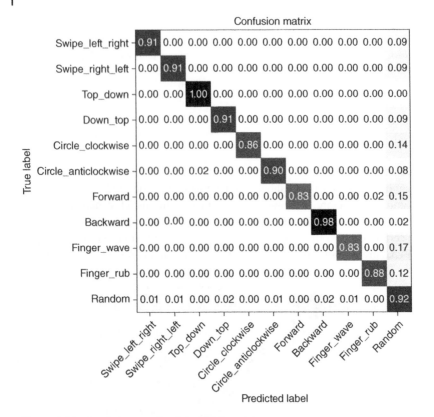

Figure 2.16 Confusion matrix of the TVAE model.

gestures is shown in both cases. That means, if a sample is recognized as a known gesture, then it is very likely classified as the right gestures. This is reasonable since a gesture is only accepted and predicted when its confidence score is high. Moreover, the last column shows how many samples of known classes are rejected. Therefore, the lower confusion values in the TVAE center CM shown the improved rejection capability of the network.

2.6.4.4 Clustering Score

The random samples recorded for the experiments in the paper are only a small extract of possible background movements that can appear in real-world scenarios. By assuming that the embedded features of background movements are uniformly distributed over the embedding space, cluster score metrics can be used to estimate the performance in real-world environments. Therefore, the silhouette score and the Davies–Bouldin score are evaluated for the embedded class clusters. The silhouette score is optimized at 1, whereas the Davies–Bouldin score reaches its

Table 2.3 Clustering scores of the class clusters in the embedding space after training.

Approach	Silhouette score	Davis–Bouldin score
Softmax	0.34 (±0.01)	1.06 (±0.03)
Triplet	0.34 (±0.02)	1.05 (±0.05)
TVAE center	0.91 (±0.01)	0.13 (±0.01)

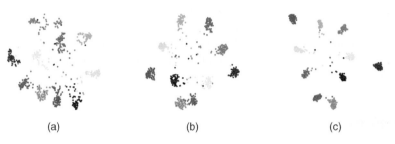

(a)   (b)   (c)

Figure 2.17 t-SNE plots of the embedded features resulting from the different models. (a) Softmax, (b) triplet, and (c) TVAE center.

best score at 0. The clustering scores of all evaluated approaches are stated in Table 2.3. Similar to the F1-scores, a static improvement of the different models can be observed. In order to illustrate the clustering results graphically, the embedded features, are projected down to two dimensions using t-stochastic neighborhood embedding (SNE) and plotted in Figure 2.17 for the softmax, triplet, and TVAE center models.

2.6.4.5 Discussion

The experiments show that assuming a closed-set environment, all approaches yield good accuracy scores. However, when showing background noise or alien motions to the network, current state-of-art networks show poor results. Nonetheless, well-considered metric learning extensions can increase the performance in open-set environments significantly. The TVAE projects the input images to a continuous distribution from which a new embedded feature vector is sampled. The sampled vector serves as basis for reconstruction, which means that it is slightly different, although it belongs to the same input image. As a result, the embedding space gets continuous, and it is guaranteed that close-by embedded feature vectors are reconstructed to the same or at least very similar spectrogram images. Especially posttraining the network using a center loss function additionally increases the compactness of class clusters in the embedding space and therefore results in an improved rejection capability.

2.7 Questions to the Reader

- What is an open-set classification problem?
- Why is the softmax normalization inappropriate to solve an open-set classification problem?
- What are the drawbacks of pairwise metric learning techniques?
- What is the advantage of using a margin compared to a continuous intra- and interclass optimization?
- What is the drawback of triplet loss with respect to intra- and interclass optimization and how does the N-pair loss target this issue?
- What is the curse of dimensionality?
- For which kind of classification problems can the LAR loss be used?
- What are the advantages of end-to-end learning techniques compared to pairwise techniques?

References

1 Yang, P., Huang, K., and Hussain, A. (2018). A review on multi-task metric learning. *Big Data Analytics* (3). https://bdataanalytics.biomedcentral.com/articles/10.1186/s41044-018-0029-9.

2 Chopra, S., Hadsell, R., and LeCun, Y. (2005). Learning a similarity metric discriminatively, with application to face verification. *2005 IEEE Computer Society Conference on Computer Vision and Pattern Recognition (CVPR'05)*, Volume 1, pp. 539–546.

3 Schroff, F., Kalenichenko, D., and Philbin, J. (2015). FaceNet: A unified embedding for face recognition and clustering. *Proceedings of the IEEE Conference on Computer Vision and Pattern Recognition (CVPR)*.

4 Chen, W., Chen, X., Zhang, J., and Huang, K. (2017). Beyond triplet loss: a deep quadruplet network for person re-identification. *2017 IEEE Conference on Computer Vision and Pattern Recognition, CVPR 2017*, Honolulu, HI, USA (21–26 July 2017). IEEE Computer Society, pp. 1320–1329. https://doi.org/10.1109/CVPR.2017.145.

5 Sohn, K. (2016). Improved deep metric learning with multi-class N-pair loss objective. In: *Advances in Neural Information Processing Systems*, vol. 29 (ed. D. Lee, M. Sugiyama, U. Luxburg et al.). Curran Associates, Inc.. https://proceedings.neurips.cc/paper/2016/file/6b180037abbebea991d8b1232f8a8ca9-Paper.pdf.

6 Rippel, O., Paluri, M., Dollar, P., and Bourdev, L. (2015). Metric learning with adaptive density discrimination. *arXiv preprint arXiv:1511.05939*.

7 Song, H., Xiang, Y., Jegelka, S., and Savarese, S. (2016). Deep metric learning via lifted structured feature embedding (2015).

8 Oh Song, H., Jegelka, S., Rathod, V., and Murphy, K. (2017). Deep metric learning via facility location. *Proceedings of the IEEE Conference on Computer Vision and Pattern Recognition*, pp. 5382–5390.

9 Chen, L. and He, Y. (2018). Dress fashionably: learn fashion collocation with deep mixed-category metric learning. *Proceedings of the AAAI Conference on Artificial Intelligence* 32 (1): 2103–2110.

10 Wang, X., Han, X., Huang, W. et al. (2019). Multi-similarity loss with general pair weighting for deep metric learning. *Proceedings of the IEEE/CVF Conference on Computer Vision and Pattern Recognition(CVPR)*.

11 KAYA, M. and BILGE, H.S. (2019). Deep metric learning: a survey. *Symmetry* 11 (9). https://www.mdpi.com/2073-8994/11/9/1066.

12 Li, Z. and Tang, J. (2015). Weakly supervised deep metric learning for community-contributed image retrieval. *IEEE Transactions on Multimedia* 17 (11): 1989–1999.

13 Yang, J., She, D., Lai, Y.-K., and Yang, M.-H. (2018). Retrieving and classifying affective images via deep metric learning. *Proceedings of the AAAI Conference on Artificial Intelligence* 32 (1): 491–498.

14 Cao, R., Zhang, Q., Zhu, J. et al. (2020). Enhancing remote sensing image retrieval using a triplet deep metric learning network. *International Journal of Remote Sensing* 41 (2): 740–751.

15 Bendale, A. and Boult, T.E. (2016). Towards open set deep networks. *Proceedings of the IEEE Conference on Computer Vision and Pattern Recognition*, pp. 1563–1572.

16 Wang, F., Xiang, X., Cheng, J., and Yuille, A.L. (2017). NormFace: L_2 hypersphere embedding for face verification. *Proceedings of the 25th ACM International Conference on Multimedia*, ser. MM '17. New York, NY, USA: Association for Computing Machinery, pp. 1041–1049. https://doi.org/10.1145/3123266.3123359.

17 Liu, W., Wen, Y., Yu, Z. et al. (2017). SphereFace: Deep hypersphere embedding for face recognition. *Proceedings of the IEEE Conference on Computer Vision and Pattern Recognition (CVPR)*.

18 Wang, H., Wang, Y., Zhou, Z. et al. (2018). CosFace: Large margin cosine loss for deep face recognition. *Proceedings of the IEEE Conference on Computer Vision and Pattern Recognition (CVPR)*.

19 Deng, J., Guo, J., Xue, N., and Zafeiriou, S. (2019). ArcFace: Additive angular margin loss for deep face recognition. *Proceedings of the IEEE/CVF Conference on Computer Vision and Pattern Recognition (CVPR)*.

20 He, L., Wang, Z., Li, Y., and Wang, S. (2020). Softmax dissection: towards understanding intra- and inter-class objective for embedding learning. *Proceedings of the AAAI Conference on Artificial Intelligence* 34 (07): 10-957–10-964. https://ojs.aaai.org/index.php/AAAI/article/view/6729.

21 Wen, Y., Zhang, K., Li, Z., and Qiao, Y. (2016). A discriminative feature learning approach for deep face recognition. In: *Computer Vision – ECCV 2016* (ed. B. Leibe, J. Matas, N. Sebe, and M. Welling), 499–515. Cham: Springer International Publishing.

22 Houle, M.E., Kriegel, H.-P., Kröger, P. et al. (2010). Can shared-neighbor distances defeat the curse of dimensionality? *International Conference on Scientific and Statistical Database Management.* Springer, pp. 482–500.

23 Zimek, A., Schubert, E., and Kriegel, H.-P. (2012). A survey on unsupervised outlier detection in high-dimensional numerical data. *Statistical Analysis and Data Mining: The ASA Data Science Journal* 5 (5): 363–387.

24 Stadelmayer, T., Stadelmayer, M., Santra, A. et al. (2020). Human activity classification using mm-Wave FMCW radar by improved representation learning. *Proceedings of the 4th ACM Workshop on Millimeter-Wave Networks and Sensing Systems,* ser. mmNets'20. New York, NY, USA: Association for Computing Machinery. https://doi.org/10.1145/3412060.3418430.

25 Movshovitz-Attias, Y., Toshev, A., Leung, T.K. et al. (2017). No fuss distance metric learning using proxies. *Proceedings of the IEEE International Conference on Computer Vision (ICCV).*

26 Kim, S., Kim, D., Cho, M., and Kwak, S. (2020). Proxy anchor loss for deep metric learning. *IEEE/CVF Conference on Computer Vision and Pattern Recognition (CVPR).*

27 Qian, Q., Shang, L., Sun, B. et al. (2019). SoftTriple loss: deep metric learning without triplet sampling. *Proceedings of the IEEE/CVF International Conference on Computer Vision (ICCV).*

28 Wan, W., Zhong, Y., Li, T., and Chen, J. (2018). Rethinking feature distribution for loss functions in image classification. *Proceedings of the IEEE Conference on Computer Vision and Pattern Recognition (CVPR).*

29 Servadei, L., Sun, H., Ott, J. et al. (2021). Label-aware ranked loss for robust people counting using automotive in-cabin radar. *arXiv preprintarXiv:2110.05876.*

30 Ge, W. (2018). Deep metric learning with hierarchical triplet loss. *Proceedings of the European Conference on Computer Vision (ECCV).*

31 Yang, Z., Bastan, M., Zhu, X. et al. (2021). Hierarchical proxy-based loss for deep metric learning. *arXiv preprint arXiv:2103.13538.*

32 Stadelmayer, T., Santra, A., Weigel, R., and Lurz, F. (2022). Radar-based gesture recognition using a variational autoencoder with deep statistical metric learning. *IEEE Transactions on Microwave Theory and Techniques.* doi: 10.1109/TMTT.2022.3201265.

3

Deep Parametric Learning

At the end of this chapter, reader will be able to learn

- How transformation-aware processing can be integrated into a neural network.
- How learned parametric layers can be customized to a specific use-case and handle specific artifacts.
- How parametric layer enables faster model convergence by limiting the optimization search space.

3.1 Introduction

Radar is a very powerful yet complex sensor. The radar signal is multidimensional, and along each dimension another physical size can be measured. Along the fast-time axis, the range is determined; along the slow-time axis, the Doppler or radial velocity is estimated; along the spatial dimensions, i.e., multiple receiving antennas, the azimuth as well as elevation angle of arrival can be determined. As derived in Chapter 1, each physical size is encoded in the frequency along the respective dimension. Thus, to extract meaningful features of the object of interest in the field of view, it is essential to estimate the frequencies with high accuracy.

Consequently, radar signals are typically transformed to frequency domain using a STFT. After STFT, the frequency components are unveiled as local peaks in the frequency domain which leads to a more image-like representation and for humans more interpretable representation. In Figure 3.1, a comparison between raw time domain radar data and its respective range-Doppler image (RDI) representation is shown. On the other side, the signal is still multidimensional and might contain complex patterns across multiple frames or RDIs. So specially for radar-based classification tasks such as human activity recognition or hand gesture recognition people proposed to use machine learning on handcrafted features

Methods and Techniques in Deep Learning: Advancements in mmWave Radar Solutions, First Edition.
Avik Santra, Souvik Hazra, Lorenzo Servadei, Thomas Stadelmayer, Michael Stephan, and Anand Dubey.
© 2023 The Institute of Electrical and Electronics Engineers, Inc. Published 2023 by John Wiley & Sons, Inc.

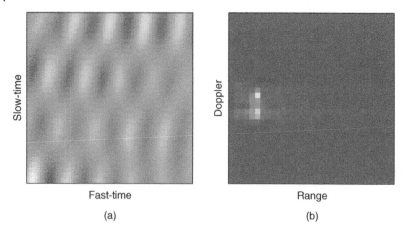

Slow-time

Doppler

Fast-time

Range

(a)

(b)

Figure 3.1 Visualization of the same radar data frame in (a) raw time domain and (b) range-Doppler domain. The radar data snapshot was taken from a finger wave hand gesture, where up and down finger movements were performed closely in front of the radar.

first [1–5]. As pointed out in [6], the classical machine learning approaches with manually extracted features work well for rather simple tasks; however, when it comes to more complex tasks, deep learning-based models show significant superior results. As a result an increasing number of papers using deep neural networks to interpret radar data were published in recent years [7–12]. Therefore, instead of doing a classification on manually extracted and thus eventually suboptimal features, deep neural networks learn to extract optimal features on its own.

For deep learning networks, typically STFT-processed data is used as input. Depending on the application, different input representations, such as range-, Doppler-, angle-spectrograms [5, 8–11, 13, 14], range-Doppler or range-angle maps [15–17] or range-Doppler-angle cubes [18], are used. Since the radar data in the frequency domain have an image-like representation, advanced neural network architectures from computer vision domain can be exploited and lead to remarkable results. On the other side, preprocessing the radar data is a manual and static processing step, which might lead to suboptimal representations for a specific task and thus might limit the overall system performance. Therefore, the question arises whether also the preprocessing can benefit from deep learning.

Naively, one can try to train the same neural network that worked well for pre-processed data also for raw or time domain radar data. However, the network will fail to extract meaningful features. The reason for this lies in the time-frequency uncertainty principle, which says that the time resolution Δt and the frequency resolution $\Delta\omega$ are related by

$$\Delta t \Delta\omega \geq c \tag{3.1}$$

where *c* is a constant. Consequently, a high-frequency resolution comes at the cost of a poor time resolution. As discussed above, the information in the radar data is encoded in the frequencies, and thus, in order to extract meaningful features, the filter kernels require a good frequency selectivity. To achieve a good frequency selectivity – according to the time-frequency uncertainty principle – a wide observation window, i.e., a long filter kernel, is required.

Even when using a simple single-input, single-output radar system with only one transmit and one receive antenna, one radar data frame is two-dimensional. Thus, two-dimensional filter kernels are required, which means that the number of trainable parameters scales quadratically with the filter length. As a result, compared to preprocessed and image-like data representations, where small filter kernels of size (3, 3) or (5, 5) can be used, raw time domain data require filter kernels in the order of magnitude of size (16, 16), (32, 32), or (64, 64) and thus lead to exploding number of trainable parameters. This in turn leads to certain problems. First of all, the memory requirements of the neural network get larger, which might be a serious concern, when it comes to embedded and memory constrained systems. Second, networks with a large number of trainable parameters tend to overfit and require more training data in order to converge to a good solution. Data acquisition is a very costly and time-consuming task which makes the data set size an important factor.

Audio signals are related to radar signals in a way that the information is also encoded in the frequency. Therefore, a similar problem of large and difficult-to-tune filter kernels when learning from time domain signals arises. As a result, researchers proposed audio-signal analysis to merge the advantages of signal processing with the power of deep learning [19–22], which showed promising results. However, not only in audio-signal processing but also in many other domains, it was successfully proposed that domain knowledge be merged into deep neural networks. The reduction of computational burden to train deep neural networks, the reduction of required data set size as well as the need for more interpretable and reliable networks were the main driving factors for using the so-called "model-based neural networks" [23].

Ravanelli and Bengio [19] proposed as one of the first to pre-initialize the filter kernels to sinc filters in order to solve an audio speaker identification task. Additionally, the training process is restricted in a way that only the filter parameters, i.e., the center frequency and bandwidth of the sinc filters, instead of each independent filter weight, is optimized. As a result of this constraint, the filter kernels in the convolutional layer can be seen as parametric kernels, since they are defined by a limited set of filter parameters. Therefore, we refer to such a constraint layer as a parametric convolutional layer. Although this restriction limits the search space and thus also the freedom of the neural network, it does it in a very efficient way. Since we have prior domain knowledge that the sinc filters extract meaningful features, we can significantly reduce the search space while still achieving perhaps not the optimal but close-to-optimal solution.

In recent years, also researchers in radar signal processing started to explore the advantages of hybrid architectures. Inspired by the SincNet [19] from 1D audio signal processing, a similar approach using parametric convolutional layers for two-dimensional input data were proposed for the first time in the radar domain in 2021 [24]. Similarly a parametric complex frequency extraction layer (CFEL) was proposed in [25]. The CFEL mimicks a DFT and allows the network only to optimize the frequencies at which the signal is sampled in frequency domain and thus can be seen as trainable nonuniform DFT layer. Similar to this, Zhao et al. proposed to introduce prior knowledge into the neural network by preinitializing the filter kernels in the first convolutional layer by the harmonic frequencies and hence mimicking a DFT [26]. However, the difference to the CFEL is that the training process is not restricted, but each filter weight is optimized independently. This simplifies the training in a way that the training starts at a reasonable level; however, it does not restrict the search space in an efficient way as the parametric layers do. Therefore, the number of trainable weights is not reduced, and after training, the filter kernels are very likely to diverge from the initial filter type, which makes them less interpretable.

In Section 3.2, we present several parametric neural network approaches specifically proposed for the radar domain. First, we introduce the radar parameteric kernels comprising of 2D sinc filters, 2D wavelet filters, and adaptive sinc filters, and then the CFEL is presented that is equivalent to a nonuniform discrete Fourier transform. In Section 3.3, a multilevel wavelet decomposition network is presented that approximates a multilevel wavelet decomposition in a neural network for time series data classification or forecasting. In Section 3.4, we present the application of activity classification using a parametric neural network namely 2D sinc filters or 2D wavelet filters and present the advantages of a parametric neural network under different situations. We then conclude the chapter summarizing the key points from the chapter followed by questions to the readers.

3.2 Radar Parametric Neural Network

As introduced in Section 3.1, a parametric neural network is a neural network with at least one layer, where the weights cannot be learned independently, but where the weights w are constraint to a certain type of function or operation that is defined by a set of parameters θ, where $\theta \ll \mathcal{N}(w)$. In this section, we discuss different parametric layers used in radar signal processing. In a traditional radar processing pipeline, an explicit feature generation, such as a range-Doppler transformation, is performed before feeding the feature images or data to a neural network

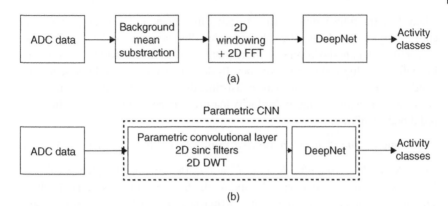

Figure 3.2 (a) Conventional processing pipeline, involving explicit preprocessing, feature generation, and neural network. (b) Proposed processing pipeline, involving 2D sinc filter parametric CNN or 2D wavelet filter parametric CNN for implicit preprocessing, feature generation, and classification.

such as a CNN or a LSTM for classification as depicted in Figure 3.2a. The 2D parametric neural network implicitly performs the preprocessing and feature generation in the neural network itself, thus the input to the parametric neural network is the raw ADC data as depicted in Figure 3.2b. The parametric layer performs a data-driven preprocessing, where the hyperparameters of a certain preprocessing operation such as a 2D sinc or wavelet filtering or a trainable nonuniform DFT are optimized according to the training data. The trainable hyperparameters for different preprocessing approaches are described in more detail later in this chapter. In some literature, this is referred to as structure-aware or transformation-aware neural network. The capability of the CNN to directly operate on the raw ADC data eliminates the need for a digital signal processor (DSP) before feeding the features into the neural network that is typically running on a specialized hardware.

3.2.1 2D Sinc Filters

Besides applying a STFT, time domain bandpass filters can be used to analyze the frequency composition of a signal. Time domain bandpass filters yield the ability to adjust the cut-off frequencies according to the specific needs of the task and can therefore be trained through data-driven optimization within a parametric neural network. Thus, extending Ravanelli and Bengio's work [19], a 2D sinc filter can be chosen as parametric filter function for radar time domain processing. The 1D sinc filter is defined as follows:

$$h_{K,f_s}(k, f_1, b) = 2(f_1 + b)\text{sinc}\left(2(f_1 + b)\frac{k - \lfloor\frac{K}{2}\rfloor}{f_s}\right) - 2f_1\text{sinc}\left(2f_1\frac{k - \lfloor\frac{K}{2}\rfloor}{f_s}\right)$$

$$(3.2)$$

where K is the filter length, f_s the sampling frequency of the signal, f_l the lower cut-off frequency, b the bandwidth, and k the filter parameter index. The parameters of this filter are the lower cut-off frequency f_l and the bandwidth b that implicitly defines the higher cut-off frequency. By defining a lower cut-off frequency and bandwidth in slow-time as well as in fast-time direction, a 2D bandpass filter that is able to extract joint range and velocity features can be created. The 2D sinc filter is defined as follows:

$$\text{sinc}_{2D}(n, m; f_1^{st}, b_{st}, f_1^{ft}, b_{ft}) = w(n, m) h_{N, f_s^{st}}(n, f_{1,st}, b_{st}) h_{M, f_s^{ft}}(m, f_{1,ft}, b_{ft}) \quad (3.3)$$

where N and M are the filter-lengths, f_s^{st} and f_s^{ft} the sampling frequencies, f_1^{st} and f_1^{ft} the lower cut-off frequencies, b_{st} and b_{ft} the filter bandwidths, respectively, in slow-time and in fast-time directions. Furthermore, $w(n, m)$ is a 2D cosine weighting function, n is sweeping along slow-time, and m along fast-time. An exemplary 2D sinc filter is shown in Figure 3.3 in time as well as in frequency domain. In frequency domain, the rectangular shape with clear cut-off frequencies can be seen. The first layer of a CNN is initialized according to the definition of 2D sinc filters and only the filter parameters are allowed to be learned during training.

3.2.2 2D Morlet Wavelets

The range-velocity profile is not only defined by its composed frequencies but also by the change of frequencies over time. When transforming a signal-to-frequency domain, the time information is lost. This can be overcome by segmenting the time domain signal into smaller chunks. However, smaller segment sizes mean higher time resolution but at the cost of frequency resolution and vice versa. Especially for time-varying signals, wavelets have the advantage over Fourier transformations, that they have different levels of time frequency resolution. Due to the fact that radar signals are highly time-varying, 2D Morlet wavelets function as parameterized kernel functions are a suitable candidate. The 2D Morlet wavelet

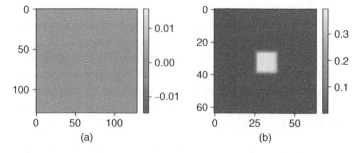

Figure 3.3 Exemplary 2D sinc filter in (a) time and (b) frequency domains.

is composed of a 2D Gaussian window and a 2D frequency and is defined as follows:

$$g_{N,M}(n, m; \sigma_{st}, \sigma_{ft})$$

$$= \frac{1}{2\pi\sigma_{st}\sigma_{ft}} \exp\left(-\frac{(n/N - \lfloor N/2 \rfloor)^2}{2\sigma_{st}^2} - \frac{(m/M - \lfloor M/2 \rfloor)^2}{2\sigma_{ft}^2}\right) \tag{3.4}$$

$$\phi[n, m; f_{st}, \sigma_{st}, f_{ft}, \sigma_{ft}]$$

$$= g_{N,M}(n, m; \sigma_{st}, \sigma_{ft}) \cos\left(2\pi f_c^{st} \frac{n - \lfloor N/2 \rfloor}{f_s^{st}}\right) \cos\left(2\pi f_c^{ft} \frac{m - \lfloor M/2 \rfloor}{f_s^{ft}}\right) \tag{3.5}$$

where N and M are the filter-lengths, σ_{st} and σ_{ft} the standard deviations, f_c^{st} and f_c^{ft} the center frequencies, f_s^{st} and f_s^{ft} the sampling frequencies, respectively, in slow-time and fast-time directions. The filter parameters that can be optimized during the training process are the center frequencies and the standard deviations of the 2D wavelet. Similar to the 2D sinc filters, the frequency area of interest can be adjusted by the center frequency, but additionally, also the time-frequency resolution can be optimized by changing the standard deviation of the Gaussian part of the wavelet function. Due to the fact that the defined wavelet is the product of a cosine and a Gaussian window function also the frequency response has the shape of a Gaussian. This results in the fact that there are no clear cut-off frequencies as it can also be seen in Figure 3.4 where an exemplary 2D Morlet wavelet in time and frequency domains is depicted. The standard deviations of the Gaussian in time domain and in frequency domain are indirectly proportional. Consequently, decreasing the width of the Gaussian in time domain will lead to an increased width of the frequency response, which illustrates the trade-off in time-frequency resolution.

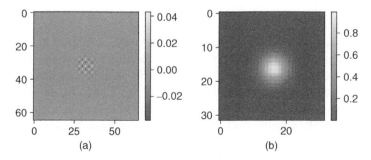

Figure 3.4 Exemplary 2D Morlet wavelet in (a) time and (b) frequency domains.

3.2.3 Adaptive 2D Sinc Filters

In the case of human target sensing in a room, there are specific range bins, which are perturbed by clutter from blinds or curtains, also based on the installation in the range-frequency, there are interference from the 60 Hz line frequencies, which can manifest as spurious targets in the radar data. Furthermore, the walking speed might change when stopping, turning around, and moving into the opposite direction and walking in an angle to radar sensor. Therefore, instead of fixed sinc filters learned once during offline training, it is often advantageous to design adaptive 2D sinc filters that follow target's movement and/or is tailored to the room it is deployed in. In order to achieve adaptive sinc filtering to faithfully extract target features and suppress interference, the fast-time center frequency and the slow-time center frequency are adapted according to the current observed signal. This can be done for each dimension independently with a similar working mechanism.

First, a complex-valued 2D sinc filter is defined with an filter length of N_s in range direction, which is equal to the number of samples per chirp and a filter length of N_c in Doppler direction, which is smaller than the number of chirps N_{cc} within a data frame. The raw signal of a data frame is convolved with this filter in range direction and also along slow-time dimension. Consequently, the 2D convolution results in a one-dimensional slow-time signal within the range frequency region that is defined by the 2D sinc filter. The same procedure is done by using a 2D sinc filter with filter length of N_c in range direction and N_s in Doppler direction.

For a data frame at time t, the average angular velocity for each complex-valued 1D signal s_t obtained after sinc filtering is determined as follows:

$$\omega_{\text{avg}}, t = \frac{1}{N-1} \sum_{n=1}^{N-1} \left[\phi(s_t[n]) - \phi(s_t[n-1]) \right] \tag{3.6}$$

where $\phi(.)$ denotes the phase. Based on the average angular velocity, the center frequency for the new time step can be updated by

$$f_c, t+1 = \frac{\omega_{\text{avg}}, t}{2\pi} \tag{3.7}$$

for slow-time and fast-time independently in the same manner. By this, the 2D sinc filter is basically scanning the target in range-Doppler domain by completely operating on the raw time domain data. In Figure 3.5, the adaptive sinc filtering approach is visualized. The RDI at three different time steps from a recording of a single person approaching the radar and walking away from it along with the frequency response of the adaptive tracking sinc filter is presented.

3.2.4 Complex Frequency Extraction Layer

Processing the data using a 2D FFT directly unveils range and Doppler information but in turn is compute intense and the number of sampling points limits its

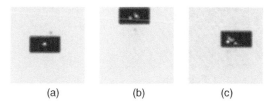

(a)　　　　　　　(b)　　　　　　　(c)

Figure 3.5 Target tracking visualized. A person is approaching the radar and moves away again. The images show the RDM together with the frequency response of the tracking sinc filter at three different time steps. (a) RDM 1, (b) RDM 2, and (c) RDM 3.

accuracy. Instead, a CFEL for range and Doppler transformation can be used. The CFEL is a parametric layer, where the weight matrix is initialized by the 2D DFT matrix and where only the sampling frequencies of the DFT can be optimized. Therefore, in the proposed CFEL, the weight matrices are defined as follows:

$$f_{M,N}(f_{ft}, f_{st}; m, n) = e^{j2\pi(mf_{ft}/f_s^{ft} + nf_{st}/f_s^{st})} \tag{3.8}$$

where m and n are the sample indices, M and N the filter lengths, and f_s^{ft} and f_s^{st} are the sampling frequencies in fast-time and slow-time, respectively. Moreover, f_{ft} and f_{st} are the trainable hyperparameters that define the weights. These hyperparameters also define the frequencies that are extracted from the signal by the CFEL. Additionally, to create a set of filter kernels, the number of filters in fast-time N_{ft}, as well as in slow-time N_{st}, has to be given. Although each filter is applied and trained independently, the output channels are reshaped to a two-dimensional matrix of size $N_{ft} \times N_{st}$ to obtain a similar representation as a RDI.

Learning the filter frequencies enables the possibility, unlike in an FFT, to analyze the signal composition in some frequency areas in more detail than in others. In order to enable equal training, the hyperparameters are normalized. When the filters are created using the set of harmonic frequencies, the output of the CFEL equals a 2D DFT, and alternately, the CFEL implements a nonuniform DFT.

The real and imaginary parts, as well as the frequency response of an exemplary filter kernel, are shown in Figure 3.6. It can be seen that the frequency response is a single sharp peak. Thus, applying a set of these filter kernels can be seen as sampling the underlying signal in frequency domain at nonuniform positions dictated by the training data and the task.

It is therefore proposed to feed raw time-domain data into the neural network and extract meaningful features using a parametric CFEL, which is the initial layer to the integrated neural network. As a result, the CFEL replaces the range-Doppler processing, which typically involves a uniform 2D DFT. As a comparison, the time domain signal, the absolute values of the preprocessed RDI, and the absolute values of the output of the already-trained CFEL of the same scene are shown in Figure 3.7. In Figure 3.7a,b, the x-axis goes along slow-time and the y-axis along fast-time, whereas in Figure 3.7c, the x- and y-axis goes along range and Doppler, respectively. It can be seen that it is hardly possible to obtain information from

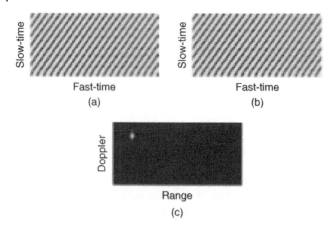

Figure 3.6 (a) Real part, (b) imaginary part, and (c) frequency response of an exemplary CFEL filter kernel.

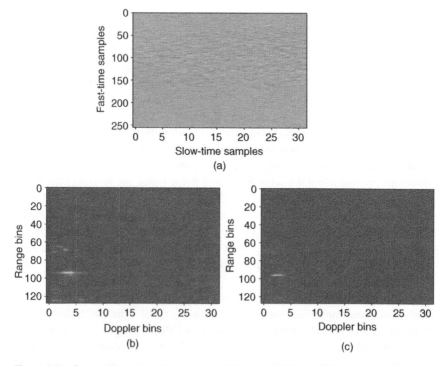

Figure 3.7 Comparison of (a) time domain ADC data, (b) RDI, and (c) corresponding output of the CFEL layer.

the time domain signal itself, but after preprocessing, the target gets unveiled as a peak in the range-Doppler domain. After initialization, the features after the CFEL layer look the same as the RDI, but during training, the frequencies that are analyzed or extracted are optimized. Thus, if the application requires, it is possible to get a higher frequency resolution in more-meaningful frequency areas and less frequency resolution in less-meaningful areas. This is an advantage, especially in classification tasks. However, if all frequency regions are of equal interest for the application, the analyzed frequencies are distributed over the whole domain. In Figure 3.7, it can be seen that in the already-trained CFEL, the target information is successfully extracted from the time domain signal. Both the preprocessed RDI and the output of the CFEL show consistent results.

3.3 Multilevel Wavelet Decomposition Network

In [27], a wavelet-based neural network structure called multilevel Wavelet Decomposition Network (mWDN) is proposed for time series analysis. Some radar applications such as vital sensing or certain implementations of gesture sensing are based on time series analysis. Therefore, the mWDN can be applied for certain radar applications even if it was not specifically designed for radar. The parameters in the multilevel discrete wavelet decomposition are learned under data-driven optimization of the neural network. Utilizing the mWDN framework, a classification framework is proposed namely the Residual Classification Flow (RCF) and additionally a forecasting framework is proposed namely the multi-frequecy Long Short-Term Memory (mLSTM). In the case of RCF, a pipelined classifier stack is used to exploit hidden features in a subseries through residual learning methods. In the case of mLSTM, the idea is to turn the hidden features in different frequency levels into inferring future states. The wavelet decomposition is approximated into a neural network framework that can be optimized via backpropagation algorithm as depicted in Figure 3.8.

The multilevel discrete wavelet decomposition (MDWD) extracts time-frequency features at multiple levels by decomposing the time series into a low- and a high-frequency subseries up to several levels. The input time series is denoted as $x = [x_1, \ldots, x_t, \ldots, x_T]$, and the low and high subseries generated in the ith level are denoted as $x^l(i)$ and $x^h(i)$. In the $(i + 1)$th level, the MDWD uses a low-pass filter $l = [l_1, \ldots, l_k]$ and a high-pass filter $h = [h_1, \ldots, h_k]$, with $k < T$, to convolute the low-frequency subseries of the upper level as follows:

$$a_n^l(i + 1) = \sum_{k=1}^{K} x_{n+k-1}^l(i) - l_k$$

$$a_n^h(i + 1) = \sum_{k=1}^{K} x_{n+k-1}^l(i) - h_k \tag{3.9}$$

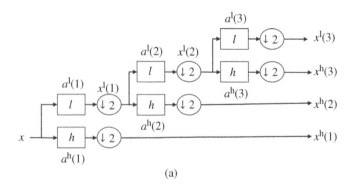

(a)

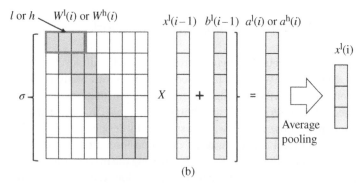

(b)

Figure 3.8 Illustration of (a) mWDN framework and (b) approximate wavelet discrete transformation. Source: Adapted from Wang et al. [27].

where $x_n^l(i)$ is the nth element of the low-frequency subseries in the ith level. The low- and high-frequency subseries $x^l(i)$ and $x^h(i)$ in the level i are generated from the $\frac{1}{2}$ down-sampling of the intermediate variable sequences $a^l(i) = [a_1^l(i), a_2^l(i), \dots]$ and $a^h(i) = [a_1^h(i), a_2^h(i), \dots]$. As the level increases, the frequency resolution is increasing, and the time resolution (for the low-frequency sub-series) is decreasing. The decomposed subseries of low and high frequencies can be fed into analogous wavelet synthesis block to reconstruct the original time series x, thus MDWD is regarded as a time-frequency decomposition.

In mWDN, the function is implemented as a neural network by approximating the MDWD transform, thus, the parameters of the low-pass filter and high-pass filter can be optimized using back-propagation. As presented in Figure 3.8, the mWDN model hierarchically decomposes a time series using the following two functions:

$$a^l(i) = \sigma(W^l(i)x^l(i-)+b^l(i))$$
$$a^h(i) = \sigma(W^h(i)x^l(i-)+b^h(i)) \tag{3.10}$$

where $\sigma(\cdot)$ is a sigmoid activation function, and $b^l(i)$ and $b^h(i)$ are trainable bias vectors, while $l(i)$ or $h(i)$ are trainable filter parameters that determine $W^l(i)$ and $W^h(i)$, respectively. The low-pass filter and high-pass filter $l(i)$ and $h(i)$ are initialized with known Wavelet filters. $x^l(i)$ and $x^h(i)$ denote the low- and high-frequency subseries of original time series x generated in the ith level, which are down-sampled using average pooling from the intermediate variables $a^l(i)$ and $a^h(i)$.

3.4 Application: Activity Classification

We use the application of activity recognition for recognizing six daily activities using FMCW radar sensor to demonstrate the advantages of using parametric neural networks. An activity dataset was recorded in a real-world environment with the radar sensor BGT60TR13C mounted on a tripod at a height of 1.20 m placed in the corner of the room. The room has about 20 m^2 with a table and chairs inside. The six daily activities are chosen in a way that it covers fast-moving activities such as walking, as well as slow-moving activities like standing idle or working on the laptop. Thus, the challenge is to cover a large Doppler velocity range and simultaneously yield a high Doppler resolution in certain regions in order to differentiate similar-looking activities in Doppler dimension. The dataset contains five different human activities plus additionally a recording of an empty room. To record the class "walking," a single human was allowed to randomly walk around. The class "idle" is split up in two recordings, including person standing in front of the radar and also person sitting idle with minor movements. The third activity is random arm movements while standing and were recorded and is referred as "arm movements." The fourth activity class is waving where the person is waving at different positions in the room facing the radar to indicate a "wake-up" kind of gesture/activity. And the fifth class, is person "working" from of the laptop-making subtle finger movements to work on the laptop. The sixth activity is "empty" denoting no person in the room. The data were acquired in a clean environment without another person or interference mainly to demonstrate the fact that parametric neural networks can differentiate activities with a wide Doppler variation while simultaneously requiring high-Doppler resolution in certain frequency bands. Each activity was performed by the same person and recorded for about 18 minutes in total.

Samples containing 2048 chirps with an overlap of 512 chirps were extracted from the recordings. Given the fact that the chirp repetition time is 1 ms each sample captures 2.048 seconds. For each sample, a Doppler spectrogram and video of RDI as described in Sections 1.5.1.1 and 1.5.1.3 is generated. Therefore, a dataset with raw ADC data, Doppler spectrograms, and video of RDI based on exactly

Table 3.1 Samples per class.

Empty room	Walking	Idle	Arm movements	Waving	Working
579	695	834	694	687	686

the same chirps per sample is obtained. For each activity, about 700 samples are available. Due to slightly different recording times for each activity, the number of samples per class varies. In Table 3.1, the exact number of samples per class is stated.

Two parametric neural networks one with 2D sinc filter kernel and 2D Morlet filter kernel are proposed to feed in raw ADC radar data as input. The parametric neural networks are compared with different the state-of-art CNN approaches. In one case, Doppler spectrogram is fed in to a 2D CNN, in second case, video of RDI is fed into a 3D CNN, and in the third case, raw ADC data are directly fed into a 2D CNN. All the network architectures have a softmax as the output layer for classifying six activities, which is trained using categorical cross-entropy loss and RMSprop is used as optimizer with a learning rate $lr = 0.0001$ and discount factor for history gradient $\rho = 0.9$ and batch size of 128. All unconstrained convolutional and dense layers are initialized using the "Glorot" initialization scheme with an uniform weight distribution. And after every common convolutional and dense layer, a dropout with a rate of 0.2 is used during training to prevent the problem of over-fitting.

3.4.1 Proposed Parametric Networks

In this section, we describe the architecture of two parametric neural networks we proposed to be used for activity classification problem.

3.4.1.1 2D SincNet

The 2D SincNet uses 2D sinc filter convolutions in the first convolutional layer as described in Section 3.2.1. As a parameter, this layer takes the filter lengths, the number of filters, the sampling frequencies, the padding mode, and the stride for the slow- as well as fast-time direction, respectively. Although there are no separated filters for slow- and fast-time, it is required to explicitly provide the number of filters in slow-time N_{st} and the number of filters in fast-time N_{ft}. According to this, 2D sinc filters are generated in a way that they form an equal grid of size $N_{st} \times N_{ft}$ covering the complete observable range Doppler domain. The trainable weights in this layer are the lower cut-off frequencies and the bandwidths in slow- and fast-time direction. In order to ensure equal significance applied to both filter dimensions, the bandwidths and cut-off frequencies are normalized.

Figure 3.9 Proposed parametric CNN learning the filter parameters of 2D sinc filters or wavelets.

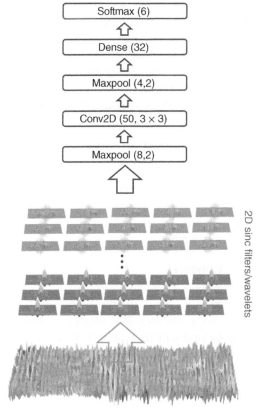

Raw 2D ADC radar data

The 2D sinc filter layer is followed by a MaxPool layer with a pooling size of 8×2, subsequently, a common two-dimensional convolutional layers using 50 filters of size 3×3 is implemented. After the dropout, a MaxPooling of size of 4×2 is applied. Then the tensor is flattened and fed into a dense layer of size 32 followed by the softmax classifier layer. The proposed network is depicted in Figure 3.9.

3.4.1.2 The 2D WaveConvNet

The 2D WaveConvNet (WCN) is designed similar to the 2D SincNet. Only the first convolutional layer is initialized by 2D Morlet wavelets as described in Section 3.2.2 instead of using 2D sinc filters. The parameters required to define this layer are the filter lengths, number of filters, sampling frequencies, padding mode, and stride for the slow- and fast-time direction, respectively. Similar to the 2D sinc filters, the number of filters in slow-time direction N_{st} and the number of filters in fast-time direction N_{ft} have to be explicitly provided in order to distribute the frequency response of the wavelets equally as a grid in the 2D frequency domain. Both time axis were normalized as in the case of 2D SincNet to ensure equal significance during training. Thus, the standard deviations is chosen to be 0.06 in both filter dimensions. In a total N_{st} times N_{ft}, 2D wavelets are defined and used in the first layer of the network. Trainable weights of this layer are the center frequencies and standard deviations in slow- and in fast-time dimension. Also, in the 2D WCN, the learnable weights are normalized.

3.4.2 The State-of-Art Networks

In this section, we describe three state-of-art neural networks, namely DSNet that takes preprocessed Doppler spectrogram as input feature, RDCNet that takes in video of RDIs as input, and 2D ConvNet that takes in raw ADC data as input; however, they do not have parametric kernels in their first layer but are rather randomly trained.

3.4.2.1 DSNet

The DSNet is a typical state-of-art 2D CNN architecture followed by a dense and softmax layers. It contains three common 2D convolutions with 4, 8, and 16 filters, respectively, and each filter uses a kernel size of 3×3. After each convolution, a dropout layer with dropout rate of 0.2 is used during training. Subsequent to the first two convolutional layers, a MaxPooling of size 2×2 is used. Afterward the tensor is flattened and fed into a dense layer of size 64 followed by the softmax classifier. The DSNet architecture is shown in Figure 3.10a.

3.4.2.2 RDCNet

The RDCNet has a temporal sequence of RDI in the form of a three-dimensional radar data cube as input. Therefore, three 3D convolutional layers are used to extract information from the radar data cube. They all have a kernel size of $3 \times 3 \times 3$ and use 4, 8, and 16 filter kernels, respectively. After the first two dropout layers, a MaxPooling of size $2 \times 2 \times 4$ is applied. Afterward the tensor is flattened and further processed by a dense layer of size 64 before it is classified by the final softmax layer. The RDCNet architecture is presented in Figure 3.10b.

3.4.2.3 The 2D ConvNet

The 2D ConvNet uses the same architecture as the 2D SincNet and 2D WCN for fair comparison. Only the first layer is substituted by a unconstrained 2D convolutional layer with "Glorot" weight initialization and no restrictions, i.e., parameterizations are applied for its kernel. Thus, each filter parameter can be learned individually dictated by the training data without any constraints.

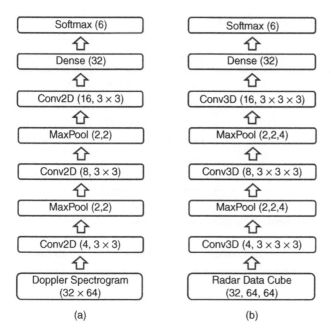

(a)   (b)

Figure 3.10 The state-of-art CNN architectures: (a) DSNet with Doppler spectrograms as input and (b) RDCNet with video of RDI as input.

The composition of parameters per layer is shown in Table 3.2 for the DSNet and the RDCNet and in Table 3.3 for the 2D SincNet, 2D WCN, and 2D ConvNet. It can be noted that the first layer of the 2D SincNet and 2D WCN have significantly less parameters compared to corresponding unconstrained convolutional layer of the 2D ConvNet. This results from the fact that only the four-filter parameter per kernel are trained in the parametric convolutional layer instead of the full kernel weight. Since 64 2D sinc filters or wavelets are used, the parametric layer has just $64 \cdot 4 = 256$ parameters to optimize. The unconstrained convolutional layer in the 2D ConvNet in contrast has to learn each single filter weight, resulting in the reduction of the network size by more than 50%.

3.4.3 Results and Discussion

The proposed approach is a data-driven preprocessing optimization. Hence, besides the evaluation on the clean dataset, the proposed approach is also evaluated regarding the impact of limited amount of training data and the impact of preknown disturbances, such as a static 50 Hz frequency of a power line, which can be considered during training.

Table 3.2 Model sizes of DSNet and RDCNet.

Layer	DSNet	Layer	RDCNet
2D Conv (4, 3 × 3)	40	3D Conv (4, 3 × 3 × 3)	112
2D Conv (8, 3 × 3)	296	3D Conv (8, 3 × 3 × 3)	872
2D Conv (16, 3 × 3)	1168	3D Conv (16, 3 × 3 × 3)	3472
Dense (32)	65 569	Dense (64)	102 432
Softmax (6)	198	Softmax (6)	198
Total	67 270	**Total**	107 086

Table 3.3 Model sizes of 2D SincNet, 2D WCN, and 2D ConvNet.

Layer	2D SincNet	2D WCN	2D ConvNet
2D Sinc (64, 65 × 33)	256	—	—
2D Wave (64, 129 × 33)	—	256	—
2D Conv (64, 65 × 33)	—	—	137 344
2D Conv (50, 3 × 3)	28 850	28 850	28 850
Dense (32)	102 432	102 432	102 432
Softmax (6)	198	198	198
Total	131 736	131 736	268 824

3.4.3.1 Clean Dataset

For evaluation, a fivefold cross-validation is performed. Thus, the model is trained five times and leaving out a different test set each time. Finally, the results of the five runs are averaged. In this way, variations in the results due to unfortunate tra7in and test data splits are reduced. All models are trained long enough to reach their saturation in training. Accordingly, the DSNet is trained for 100 epochs, the RDCNet is trained for 50 epochs, the 2D ConvNet is trained for 40 epochs, and the proposed 2D SincNet and 2D WCN are trained for 20 epochs. Afterward, the accuracy as well as the F1 score are evaluated on the test dataset. The obtained accuracies and F1 scores are averaged over all runs. Additionally, the standard deviation is calculated for both metrics.

The results are shown in Table 3.4, while the state-of-art approaches achieve an accuracy of about 90.7% and 95.6%, and 98.2%, respectively, the proposed approaches show an improved accuracy of 99.2% and 99.5%, respectively. In order

Table 3.4 Accuracies (in %) and F1-scores (in %) for the evaluated approaches on the clean dataset.

Model	Accuracy (deviation)	F1-score
DSNet	90.7 (±3.2)	90.8 (±3.0)
RDCNet	95.6 (±0.7)	95.9 (±0.6)
2D ConvNet	98.2 (±2.2)	98.2 (±2.3)
2D SincNet	99.2 (±0.6)	99.2 (±0.6)
2D WCN	99.5 (±0.3)	99.5 (±0.3)

to analyze the classification results in more detail, the confusion matrices of RDCNet representing the state-of-art approaches and the confusion matrix of 2D WCN representing the novel architectures are shown in Figure 3.11. The limitation of the state-of-art approaches is clearly observed in the fact that the state-of-art approaches are unable to differentiate between similar activities such as "idle" and "working." However, the proposed parametric NN approach is capable of distinguishing between these activities, therefore, achieving better accuracy scores.

In Figure 3.12, the validation accuracy over training epochs for the 2D ConvNet in comparison with the 2D WCN is plotted. Both models were trained five times. The vertical lines visualize the lowest as well as highest accuracy achieved by the different training runs at a certain epoch and thus represent the dispersion of the accuracies in a certain epoch. As seen in the plot for the 2D ConvNet, the mean accuracy as well as the dispersion does not improve anymore from about epoch 30 onward, which indicates that the training has already converged. On the other hand, nonparametric ConvNet approach exhibits high variation and low stability during training. This is due to the fact that the optimization problem is very high dimensional and thus has many local minima toward which the solution could be stuck. Furthermore, the learned kernel in unconstrained ConvNet can have very specific frequency selectivity, thus small changes in these filters during training induce a totally different frequency selectivity causing a large impact on the classification score. However, the proposed parametric kernels are chosen to have a controlled and continuous frequency selectivity resulting in very stable training. This can be attributed to the fact that the search space is a much lower-dimensional space parameterized by the sinc or wavelet functions and also the tractable mathematical function dictated by the sinc or wavelet function.

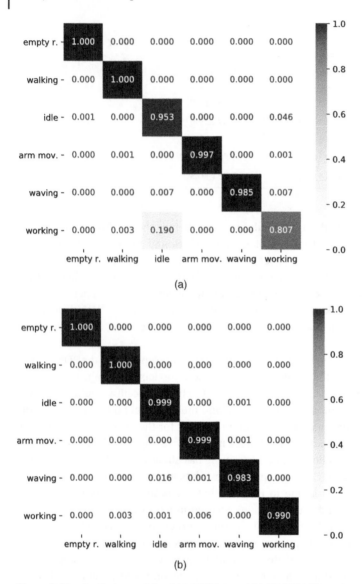

Figure 3.11 Confusion matrix of (a) RDCNet and (b) 2D WCN. The values are rounded to three decimals.

The sinc filters as well as the wavelets are initialized as described in Sections 3.4.1.1 and 3.4.1.2, while the unconstrained convolutional layer of the 2D ConvNet is initialized using "Glorot" scheme. For each approach, cumulating the initial filters leads to an approximately uniform range and velocity gain over the whole space. During training, the filter parameters are iteratively

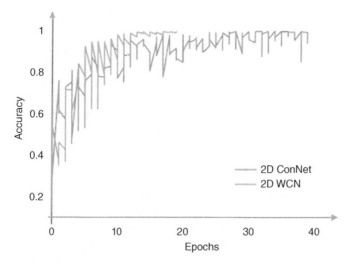

Figure 3.12 Validation accuracy over training epochs of five training runs of the 2D ConvNet and 2D WCN. Vertical lines indicate the lowest and highest achieved accuracy in a certain epoch.

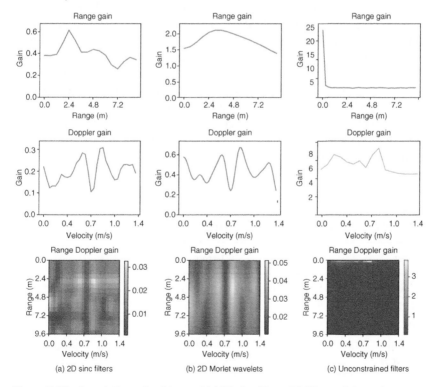

Figure 3.13 Cumulative gain of learned (a) 2D sinc filters, (b) 2D wavelets, and (c) unconstrained filters.

optimized. As a result, the initial grid structure is dissolved by individual shifts and shape changes of the filters.

The cumulative gain of all filters after training is depicted in Figure 3.13. The 2D sinc filters as well as the 2D wavelets have a bandpass characteristic and also leads to an interpretable model. Therefore, the resulting gain of cumulative filters look similar except the fact that the 2D wavelet gain is smoother caused by its smooth filter shape in frequency domain. However, the resulting weights of unconstrained convolutional layer are quite different and cannot be interpreted.

3.4.3.2 Fixed Disturbance

If there are fixed disturbances depending on the system installation such as 50 Hz interference from main power grid or a Wi-Fi signal, they can be better accounted for in the parametric neural network by creating specific nulls at those interfering frequencies. To illustrate the effect, a static 50 Hz sinusoidal signal with −12 dB with respect to the maximum detectable signal power of the sensor was added to the dataset. The networks were then retrained using the modified dataset.

In Table 3.5, the final accuracies and F1-scores are shown. The models using fixed preprocessing are effected by the disturbance. Since the 50 Hz interference is almost static within a chirp, its influence is reduced when subtracting the mean of the signal before calculating the Doppler spectrogram. The architectures operating directly on the raw ADC data have to learn to suppress the disturbing frequency. The results show that the constrained networks, namely 2D WCN and 2D Sinc-Net, perform better in suppressing the disturbance. In Figure 3.14, the adaption of the sinc filters and wavelets, respectively, to suppress the 50 Hz interference, which corresponds to a velocity of 0.25 m/s, are depicted. The learned filters of the unconstrained 2D ConvNet aren't interpretable, while the 2D SincNet and 2D WCN are interpretable.

Table 3.5 Accuracies (in %) and F1-scores (in %) under the presence of a static 50 Hz interference.

Model	Accuracy	F1-score
DSNet	88.3 (±2.7)	88.3 (±3.7)
RDCNet	95.5 (±0.6)	95.7 (±0.6)
2D ConvNet	93.2 (±4.2)	93.5 (±3.8)
2D SincNet	98.8 (±0.5)	99.5 (±0.5)
2D WCN	98.8 (±1.3)	98.9 (±1.3)

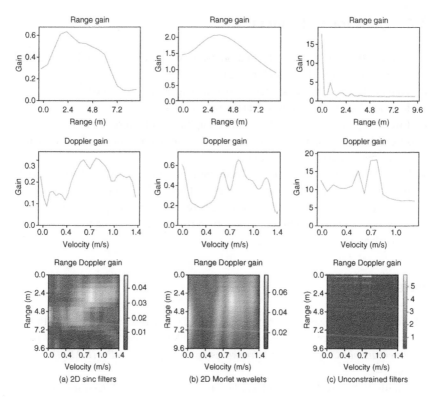

Figure 3.14 Cumulative gain of learned (a) 2D SincNet, (b) 2D WCN, and (c) unconstrained kernel in 2D CNN in the presence of a static 50 Hz interference.

3.5 Conclusion

Parametric neural networks not only help in making the neural network interpretable but also helps in improving classification accuracy in difficult cases, such as not enough separability in input features among classes or under interference. The limitation of the state-of-art approaches has its origin in preprocessing using Fourier transform for generating RDIs equally discretizes the range as well as velocity dimension. While unconstrained CNNs with raw ADC data as input are capable of learning the weights of the network, they are usually intractable and the high-dimension optimization leads to poor convergence and often overfitting. On the other hand, using parametric functions to define the kernels in a parametric neural network help in model interpretability while also rejecting interference in the data and can distinguish between similar classes via data-driven optimization.

3.6 Question to Readers

- What are the different transformation-aware processing that can be integrated into a neural network model for radar data?
- What are the specific advantages of parametric neural networks?
- How can the CFEL be compared to standard FFT?
- How does parametric neural networks help in model interpretability?
- How is the search space in case of parametric neural network limited compared to unconstrained neural network?
- What are the problems that the adaptive parametric neural network can alleviate compared to nonadaptive parametric neural network design?
- Why does parametric neural networks converge faster during training?
- Define any other parametric function that can be integrated into a neural network and specify a use-case.

References

1 Sun, Y., Fei, T., Schliep, F., and Pohl, N. (2018). Gesture classification with handcrafted micro-Doppler features using a FMCW radar. *2018 IEEE MTT-S International Conference on Microwaves for Intelligent Mobility (ICMIM)*, pp. 1–4.

2 Smith, K.A., Csech, C., Murdoch, D., and Shaker, G. (2018). Gesture recognition using mm-Wave sensor for human–car interface. *IEEE Sensors Letters* 2 (2): 1–4.

3 Kim, Y. and Ling, H. (2009). Human activity classification based on micro-Doppler signatures using a support vector machine. *IEEE Transactions on Geoscience and Remote Sensing* 47 (5): 1328–1337.

4 Bryan, J. and Kim, Y. (2010). Classification of human activities on UWB radar using a support vector machine. *2010 IEEE Antennas and Propagation Society International Symposium*, pp. 1–4.

5 Wang, Y., Ren, A., Zhou, M. et al. (2020). A novel detection and recognition method for continuous hand gesture using FMCW radar. *IEEE Access* 8: 167-264–167-275.

6 Gurbuz, S.Z. and Amin, M.G. (2019). Radar-based human-motion recognition with deep learning: promising applications for indoor monitoring. *IEEE Signal Processing Magazine* 36 (4): 16–28.

7 Zhang, Z., Tian, Z., and Zhou, M. (2018). Latern: Dynamic continuous hand gesture recognition using FMCW radar sensor. *IEEE Sensors Journal* 18 (8): 3278–3289.

8 Choi, J.-W., Ryu, S.-J., and Kim, J.-H. (2019). Short-range radar based real-time hand gesture recognition using LSTM encoder. *IEEE Access* 7: 33-610–33-618.

9 Chmurski, M., Zubert, M., Bierzynski, K., and Santra, A. (2021). Analysis of edge-optimized deep learning classifiers for radar-based gesture recognition. *IEEE Access* 9: 74-406–74-421.

10 Yu, M., Kim, N., Jung, Y., and Lee, S. (2020). A frame detection method for real-time hand gesture recognition systems using CW-radar. *Sensors* 20 (8): https://www.mdpi.com/1424-8220/20/8/2321.

11 Hernangomez, R., Santra, A., and Stanczak, S. (2019). Human activity classification with frequency modulated continuous wave radar using deep convolutional neural networks. *2019 International Radar Conference (RADAR)*, pp. 1–6.

12 Wang, S., Song, J., Lien, J. et al. (2016). Interacting with soli: exploring fine-grained dynamic gesture recognition in the radio-frequency spectrum. *Proceedings of the 29th Annual Symposium on User Interface Software and Technology*, ser. UIST '16. New York, NY, USA: Association for Computing Machinery, p. 851–860. https://doi.org/10.1145/2984511.2984565.

13 Stadelmayer, T., Santra, A., Stadelmayer, M. et al. (2021). Improved target detection and feature extraction using a complex-valued adaptive sinc filter on radar time domain data. *2021 29th European Signal Processing Conference (EUSIPCO)*, pp. 1745–1749.

14 Wang, Y., Zhao, Z., Zhou, M., and Wu, J. (2019). Two dimensional parameters based hand gesture recognition algorithm for FMCW radar systems. In: *Wireless and Satellite Systems* (ed. M. Jia, Q. Guo, and W. Meng), 226–234. Cham: Springer International Publishing.

15 Hazra, S. and Santra, A. (2018). Robust gesture recognition using millimetric-wave radar system. *IEEE Sensors Letters* 2 (4): 1–4.

16 Lei, W., Jiang, X., Xu, L. et al. (2020). Continuous gesture recognition based on time sequence fusion using MIMO radar sensor and deep learning. *Electronics* 9 (5): https://www.mdpi.com/2079-9292/9/5/869.

17 Yu, J.-T., Yen, L., and Tseng, P.-H. (2020). mmWave radar-based hand gesture recognition using range-angle image. *2020 IEEE 91st Vehicular Technology Conference (VTC2020-Spring)*, pp. 1–5.

18 Hazra, S. and Santra, A. (2019). Radar gesture recognition system in presence of interference using self-attention neural network. *2019 18th IEEE International Conference On Machine Learning And Applications (ICMLA)*, pp. 1409–1414.

19 Ravanelli, M. and Bengio, Y. (2018). Speaker recognition from raw waveform with SincNet. *2018 IEEE Spoken Language Technology Workshop (SLT)*, pp. 1021–1028.

20 Gaetan, F. and Olga, F. (2022). Learnable wavelet packet transform for data-adapted spectrograms. *arXiv preprint arXiv:2201.11069.*

21 Sailor, H.B., Agrawal, D.M., and Patil, H.A. (2017). Unsupervised filterbank learning using convolutional restricted Boltzmann machine for environmental sound classification. *Interspeech* 8: 9.

22 Xiong, S., Zhou, H., He, S. et al. (2020). A novel end-to-end fault diagnosis approach for rolling bearings by integrating wavelet packet transform into convolutional neural network structures. *Sensors* 20 (17): 4965.

23 Shlezinger, N., Whang, J., Eldar, Y.C., and Dimakis, A.G. (2020). Model-based deep learning. *arXiv preprint arXiv:2012.08405.*

24 Stadelmayer, T., Santra, A., Weigel, R., and Lurz, F. Parametric convolutional neural network for radar-based human activity classification using raw ADC data. (2021). *IEEE Sensors Journal* 21 (17): 19529–19540.

25 Stephan, M., Stadelmayer, T., Santra, A. et al. (2021). Radar image reconstruction from raw ADC data using parametric variational autoencoder with domain adaptation. *2020 25th International Conference on Pattern Recognition (ICPR)*, pp. 9529–9536.

26 Zhao, P., Lu, C.X., Wang, B. et al. (2021). CubeLearn: End-to-end learning for human motion recognition from raw mmWave radar signals. *arXiv preprint arXiv:2111.03976.*

27 Wang, J., Wang, Z., Li, J., and Wu, J. (2018). Multilevel wavelet decomposition network for interpretable time series analysis. *Proceedings of the 24th ACM SIGKDD International Conference on Knowledge Discovery & Data Mining*, pp. 2437–2446.

4

Deep Reinforcement Learning

After reading this chapter, you should be able to

- Understand the basics of reinforcement learning.
- Obtain an overview of different typologies of reinforcement learning algorithms.
- Acknowledge that continuous deep reinforcement learning can improve the tracking-parameters selection in radar sensors.

4.1 Useful Notation and Equations

In this section, we collect a set of notation and useful equations, which support the reader in understanding the concepts and use-cases presented in this chapter.

4.1.1 Markov Decision Process

S_t state at time t
A_t action at time t
R_t reward at time t
γ discount rate (where $0 \leq \gamma \leq 1$)
G_t discounted return at time t ($\sum_{k=0}^{\infty} \gamma^k R_{t+k+1}$)
S set of all nonterminal states
S^+ set of all states (including terminal states)
$\mathcal{A}$ set of all actions
$\mathcal{A}(s)$ set of all actions available in state s
$\mathcal{R}$ set of all rewards
$p(s', r|s, a)$ probability of next state s' and reward r, given current state s and current action a ($\mathbb{P}(S_{t+1} = s', R_{t+1} = r|S_t = s, A_t = a)$)

Methods and Techniques in Deep Learning: Advancements in mmWave Radar Solutions, First Edition.
Avik Santra, Souvik Hazra, Lorenzo Servadei, Thomas Stadelmayer, Michael Stephan, and Anand Dubey.

4.1.2   Solving the Markov Decision Process

π policy

> *if deterministic*: $\pi(s) \in \mathcal{A}(s)$ for all $s \in S$
>
> *if stochastic*: $\pi(a|s) = \mathbb{P}(A_t = a|S_t = s)$ for all $s \in S$ and $a \in \mathcal{A}(s)$

v_π state-value function for policy π ($v_\pi(s) \doteq \mathbb{E}[G_t|S_t = s]$ for all $s \in S$)

q_π action-value function for policy π ($q_\pi(s, a) \doteq \mathbb{E}[G_t|S_t = s, A_t = a]$ for all $s \in S$ and $a \in \mathcal{A}(s)$)

v_* optimal state-value function ($v_*(s) \doteq \max_\pi v_\pi(s)$ for all $s \in S$)

q_* optimal action-value function ($q_*(s, a) \doteq \max_\pi q_\pi(s, a)$ for all $s \in S$ and $a \in \mathcal{A}(s)$)

4.1.3   Bellman Equations

4.1.3.1   Expectation Equations

$$v_\pi(s) = \sum_{a \in \mathcal{A}(s)} \pi(a|s) \sum_{s' \in S, r \in R} p(s', r|s, a)(r + \gamma v_\pi(s'))$$

$$q_\pi(s, a) = \sum_{s' \in S, r \in R} p(s', r|s, a)\left(r + \gamma \sum_{a' \in \mathcal{A}(s')} \pi(a'|s')q_\pi(s', a')\right)$$

4.1.3.2   Optimality Equations

$$v_*(s) = \max_{a \in \mathcal{A}(s)} \sum_{s' \in S, r \in R} p(s', r|s, a)(r + \gamma v_*(s'))$$

$$q_*(s, a) = \sum_{s' \in S, r \in R} p(s', r|s, a)(r + \gamma \max_{a' \in \mathcal{A}(s')} q_*(s', a'))$$

4.2   Introduction

Reinforcement learning (RL) is a machine learning (ML) paradigm where the learning signal is not fixed. Instead, it depends on the sum of rewards produced by a policy interacting with an environment [1]. A policy interacts with an environment by choosing actions given a state, and the environment reacts by producing a new state and a reward. The objective is to find a policy which maximizes the sum of rewards over an episode, the return. Deep reinforcement learning (DRL) uses modern neural networks and deep learning (DL) techniques to generalize over states and actions. Recently, DRL has gained significant attention due to its remarkable success, mostly in game problems [2]. Deep Q-networks (DQNs) was one of the first DRL successes and instantly became one of the milestones in the area [3]. Two of the key ingredients were the usage of a target network to stabilize learning and avoid oscillations in the updates and the replay buffer to store past experiences to learn from. Many other improvements followed these two ideas. Double Q-learning [4] and dueling networks [5] are examples of how to boost learning by reducing overestimation and promoting generalization. Learning from demonstrations and imitation learning [6] was already known in robotics and the RL fields but got revisited and pushed forward in recent years. Nonetheless, when dealing with real-world problems, DRL still faces many challenges. Very often, actions in real-world problems are not discrete but continuous. Additionally, actions may be composed of many parameters making the action space combinatorial. To this end, actor-critic DRL methods as well as Q-learning methods, originally applied to discrete action spaces, have been readapted to continuous action spaces.

In order to show how the deep reinforcement learning (DRL) can help in real-life problems related to short-range radar sensors, in this chapter, we review first fundamentals of RL. Afterward, the difference between different typologies of RL algorithms is explained (e.g., On-Policy/Off-Policy, Model-based/Model-free). Finally, a real-life use-case is presented, which includes a continuous RL-based optimization for tracking-parameters of short-range radars.

RL [1] is an important branch of *ML*. RL is used to learn a decision policy from interactions with a given environment. RL algorithms learn to regulate the decision policy through trial and error, where the feedback is given by the environment in form of reward signals. The policy learning process is subdivided into episodes, which end once the terminal state of the simulation is reached. RL problems can be formulated as *Markov decision processes* (MDPs) [1]. A MDP is a tuple $< S, A, P, R, \gamma >$, where

- S is a finite set of states,
- A is a finite set of actions,
- P is a state transition probability matrix, i.e.,

$$P_{ss'}^a = \mathbb{P}[S_{t+1} = s' | S_t = s, A_t = a]$$

where $P_{ss'}^a$ is the probability that action a in state s at time t will lead to state s' at time $t + 1$,

- R is a reward function,

$$R_s^a = \mathbb{E}[R_{t+1} | S_t = s, A_t = a]$$

where R_s^a is the expectation of the immediate reward, by taking action a in state s at time t, and

- γ is a discount factor for future rewards, $\gamma \in [0, 1]$,

A policy π is a measurable mapping $\pi : S \rightarrow A$ and defines the learning agent's way of behaving at a given timestep. At every timestep, the agent experiences a reward R_{t+1} after executing an action A_t in a state S_t. The sum of all rewards collected in an episode $G_0 = \sum_{t=0}^{\infty} \gamma^t R_{t+1}$ is called the return. The goal of RL is to find an optimal policy π^* which maximizes the future expected return, i.e.,

$$\pi^* = \underset{\pi}{\operatorname{argmax}} \, \mathbb{E}_{\pi}[G_0]$$

The defining property of MDPs is the Markov property, which says that the value of the next state S_{t+1} and reward R_{t+1} of the process only depends on the actual state S_t and action A_t and not on the sequence of past events $(S_{0:t}, A_{0:t})$, i.e.,

$$\mathbb{P}[S_{t+1} = s', R_{t+1} = r \mid S_{0:t} = s_{0:t}, A_{0:t} = a_{0:t}]$$
$$= \mathbb{P}[S_{t+1} = s', R_{t+1} = r \mid S_t = s_t, A_t = a_t] \tag{4.1}$$

Moreover, in order to strengthen the RL approach with powerful function approximation capabilities, researchers investigated the combination of RL with *DL* [7] algorithms. This combination takes the name of *DRL*. DL is a class of ML algorithms which is based on *artificial neural networks* (ANNs). As explained in [7], DL is capable of processing complex and high-dimensional data for many tasks such as image classification and object detection. This capability turns out to be extremely useful, in general, once the relationship between input and

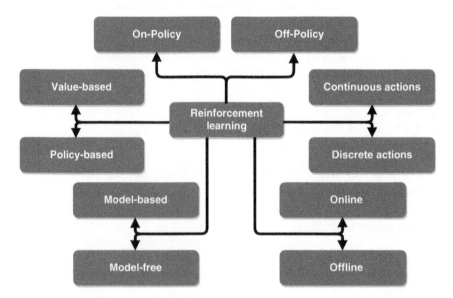

Figure 4.1 Taxonomy of the RL algorithms.

output of the task considered is highly complex. This is the case for many real-life scenarios such as applications in robotics, chip design, and autonomous driving.

Given the characteristics of the algorithms, and for the purpose of the chapter, DRL algorithms can be distinguished in the following:

- On-Policy/Off-Policy,
- Model-based/Model-free,
- Value-based/Policy-based,
- Online/Offline,
- Discrete actions-based/continuous actions-based.

The proposed taxonomy is visualized in Figure 4.1

In the next sections, we will first review these main categories and, successively, we are going to present a real-life application of DRL for short-range radars.

4.3 On-Policy Reinforcement Learning

On-Policy RL algorithms are the algorithms that evaluate and improve the same policy which is being used to select actions. In the common flow of On-Policy algorithms, a simple/soft policy is initialized. Afterward, the state-space is sampled with that policy. Finally, the policy is improved. Some examples of On-Policy algorithms are policy iteration, value iteration, Monte Carlo for On-Policy, state–action–reward–state–action SARSA, as explained in [8].

4.4 Off-Policy Reinforcement Learning

In Off-Policy learning, the agent can learn about many policies that are different from the policy being executed. Methods capable of Off-Policy learning have several important advantages over On-Policy methods. Most importantly, Off-Policy methods allow an agent to learn about many different policies at once, forming the basis for a predictive understanding of an agent's environment and enabling the learning of options [1]. Some examples of Off-Policy algorithms are Q-learning, deep deterministic policy gradient (DDPG), as described in [8].

4.5 Model-Based Reinforcement Learning

The term *model* identifies the transition probability distribution and reward function which helps define the whole environment.

Model-based RL does estimate the MDP from data. In this way, given sufficient data, the whole tuple $(S, A, \hat{R}, \hat{S}', \gamma)$ can be estimated, by inferring $\hat{R}, \hat{S}'$ from enough data.

In this setting, the MDP will converge given enough data sample. In this case, once the MDP is constructed, it can be used for solving for the optimal policy. If the state space is large, nevertheless, oftenwise the model-based approach would not converge and would not help to find the optimal policy.

4.6 Model-Free Reinforcement Learning

Model-free RL, differently from model-based approaches, does not consider to solve the MDP and so have an explicit estimation of the state transition probability, as well as the reward. To this end, the approach followed is a trial-and-error mechanism, where the feedback is given by the environment as reward signals. Being model-free RL more effective for large, complex, and highly nonlinear problems, it has been adopted more and more in the scientific literature and real-world applications.

4.7 Value-Based Reinforcement Learning

Value-based DRL is taking into account the optimization of the value (or the Q-value) function. Following the Bellman equation, which breaks down a dynamic optimization problem into a sequence of simpler subproblems, typically the value function and Q-value function are optimized by iterative RL algorithms.

The value function identifies how good is a state, which is the expected cumulative reward from a state, following the policy π. Eq. (4.4) shows how the value function is computed.

$$v_\pi(s) = \sum_{a \in A(s)} \pi(a|s) \sum_{s' \in S, r \in R} p(s', r|s, a)(r + \gamma v_\pi(s')) \tag{4.2}$$

RL algorithms which compute a Bellman optimality equation on the value function, pointed out as $v_*(s)$, are iteratively computing the optimal expected cumulative reward from a specific state. This formulation is shown in Eq. (4.3).

$$v_*(s) = \max_{a \in A(s)} \sum_{s' \in S, r \in R} p(s', r|s, a)(r + \gamma v_*(s')) \tag{4.3}$$

Once the value function is computed for both state and action, namely a state–action pair, it takes the name of Q-value function. Similarly to the value function, this expresses the expected cumulative reward, following a certain policy π from a certain state and action pair. The Q-value function is computed as follows:

$$q_\pi(s,a) = \sum_{s' \in S, r \in R} p(s', r|s, a)(r + \gamma \sum_{a' \in \mathcal{A}(s')} \pi(a'|s')q_\pi(s', a')) \tag{4.4}$$

Algorithms which propose iterations toward the optimal Q-value function, namely finding the optimal state–action pair at each time step, aim to converge toward the Bellman optimality equation expressed in Eq. (4.5):

$$q_*(s,a) = \sum_{s' \in S, r \in R} p(s', r|s, a)(r + \gamma \max_{a' \in \mathcal{A}(s')} q_*(s', a')) \tag{4.5}$$

Methods that iterate toward the optimal Q-value function $q_*(s, a)$ are obtaining the optimal policy π^* as shown in Eq. (4.6):

$$\pi^* = \underset{a \in A}{\operatorname{argmax}} \hat{Q}(s, a) \tag{4.6}$$

4.8 Policy-based Reinforcement Learning

An alternative to learning the value (or the Q-value function) function and iterate over it is policy-based algorithms. Here, the policy is learned directly from trajectories (usually stored in form of $\tau_{t:t+1} = (s_t, a_t, r_t, s_{t+1})$) and out of a set of policies. In case of neural networks-based algorithms, the policy is generally expressed by the parameters θ of a neural network, and it is used for determining the value (Q-value) of a state (state-action pair). The case of the value function is shown in Eq. (4.7).

$$V_\theta(s) = \mathbb{E}_{\pi_\theta} \left[\sum_{t=1}^{\infty} \gamma^t r_t | s' \sim p(s'|s_t, \pi_\theta(s_t)), s_1 = s \right] \tag{4.7}$$

Therefore, the optimal policy parameters are identified from Eq. (4.8):

$$J_\theta = \mathbb{E}_{\pi_\theta} \sum [\gamma r]$$
$$\theta^*(s) = \underset{\theta}{\operatorname{argmax}} J_\theta \tag{4.8}$$

Those RL methods which search directly for the optimal policy are called policy-based.

4.9 Online Reinforcement Learning

Online and offline learning methods are not specific of the RL area. Nevertheless, similar to other ML methods, RL algorithms also can be trained in an online or offline fashion.

To this end, online learning RL works with data in an incremental way. In fact, while the algorithm is deployed, data are collected continuously and the performance of the RL algorithm is improved constantly. What is usually wished in this setting is to avoid catastrophic forgetting and accumulating new knowledge without loss of previous tasks performance.

4.10 Offline Reinforcement Learning

On the other hand, offline RL is the commonly used setting, where data are collected in datasets, and the training is performed on previously gathered instances. This framework helps the reproducibility of results, as well as the learning from samples of the entire distribution of data. As a consequence, offline RL is less prone to suffering from catastrophically forgetting issues, as in online RL.

4.11 Reinforcement Learning with Discrete Actions

A set of problems which can be solved with RL entail discrete actions spaces. Having a finite set of actions usually helps in the computation and faster convergence of the problem. Additionally, many discrete action spaces problems (even with a large set of actions) are to be found in many real-life applications. These go from recommender systems, industrial plants, and language models.

4.12 Reinforcement Learning with Continuous Actions

Nevertheless, particularly in the field of robotics, autonomous driving, navigation, the actions space (and, oftentimes, the state space) is continuous. Here, usually the convergence is more difficult as the choice of actions is not finite. This generally increases the computation overhead of the RL methods, when solving for the optimal action.

4.13 Reinforcement Learning Algorithms for Radar Applications

Different typologies of RL algorithms have been applied to radar-based tasks. Using the taxonomy explained in the current section, we analyze how those algorithms have been utilized in the literature.

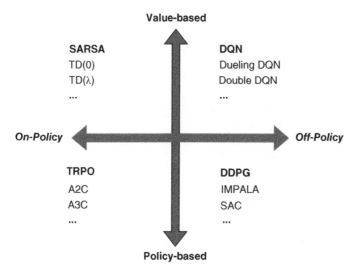

Figure 4.2 Taxonomy of the reinforcement learning algorithms.

Figure 4.2 shows the category assignment for some of the most renowned model-free RL algorithms, based on On-Policy/Off-Policy, and value-based/policy-based coordinates.

SARSA is an On-Policy, value-based method in which the Q-value is updated by using, for evaluating the action, the same policy that determined the action itself (Algorithm 4.1) [9]. The algorithm pseudocode has been shown in Algorithm 4.2. This method has been used in radar applications, for obtaining waveform optimization for multiple-input multiple-output (MIMO) radar multitarget detection, as shown in this work [10].[1] Here, a MIMO radar synthesizes and optimizes a set of transmitted waveforms, using RL, for a different number of targets and angle positions.

DQN is an Off-Policy, value-based method that combines deep Q-learning with deep neural networks. In DQN, differently from SARSA, the estimate of the optimal Q-value is obtained utilizing the maximum of the reward a the next time step s_{t+1} (and estimated by a neural network), as explained in this work [11]. The algorithm is visualized in Algorithm 4.3.[2] The algorithm is used, for instance, in this contribution [12]. Here, DQN has been used for radar-based detection and tracking in congested spectral environments. Using the DQN algorithm indeed, the radar learns to vary the bandwidth and center frequency of its

---

1 Algorithm taken from https://github.com/MartinThoma/LaTeX-examples/tree/master/ source-code/Pseudocode
2 Algorithm adapted from https://github.com/MartinThoma/LaTeX-examples/tree/master/ source-code/Pseudocode/q-learning

---

Algorithm 4.1 SARSA: Learn function $Q : \mathcal{X} \times \mathcal{A} \to \mathbb{R}$.

Require:
    Sates $\mathcal{X} = \{1, \dots, n_x\}$
    Actions $\mathcal{A} = \{1, \dots, n_a\}, \quad A : \mathcal{X} \Rightarrow \mathcal{A}$
    Reward function $R : \mathcal{X} \times \mathcal{A} \to \mathbb{R}$
    Black-box (probabilistic) transition function $T : \mathcal{X} \times \mathcal{A} \to \mathcal{X}$
    Learning rate $\alpha \in [0, 1]$, typically $\alpha = 0.1$
    Discounting factor $\gamma \in [0, 1]$
    $\lambda \in [0, 1]$: Trade-off between TD and MC
1: **procedure** SARSA($\mathcal{X}, A, R, T, \alpha, \gamma, \lambda$)
2:     Initialize $Q : \mathcal{X} \times \mathcal{A} \to \mathbb{R}$ arbitrarily
3:     **while** Q is not converged **do**
4:         Select $(s, a) \in \mathcal{X} \times \mathcal{A}$ arbitrarily
5:         **while** s is not terminal **do**
6:             $r \leftarrow R(s, a)$ ▷ Receive the reward
7:             $s' \leftarrow T(s, a)$ ▷ Receive the new state
8:             Calculate π based on Q (e.g., epsilon-greedy)
9:             $a' \leftarrow \pi(s')$
10:           $Q(s, a) \leftarrow (1 - \alpha) \cdot Q(s, a) + \alpha \cdot (r + \gamma Q(s', a'))$
11:           $s \leftarrow s'$
12:           $a \leftarrow a'$
        return Q

---

linear frequency modulated (LFM) waveform and improving the signal-to-noise ratio (SINR).

Trust region policy optimization (TRPO) is an On-Policy, policy-based method. TRPO is a policy gradient method which avoids parameter updates that change the policy too much (with regards to the Kullback–Leibler divergence) at each policy update iteration, as explained in this contribution [13]. This method has been used in this work [14], where new strategies of solving main lobe jamming have been elaborated in frequency-agile radars. The robustness of the strategies has been benchmarked against different forms of jamming perturbations.

Finally, *DDPG* is an Off-Policy, policy-based method, explained in this paper [15]. DDPG is an algorithm which concurrently learns a Q-function and a policy. It first learns the Q-function using Off-Policy data and the Bellman equation; afterward, it uses the Q-function to learn the policy. The algorithm is shown in Algorithm 4.4.[3] In this work [16], DDPG is used for solving a radar waveform design problem. The problem consists in selecting phases for a phase-coded

---

3 Algorithm taken from https://github.com/harryzhangOG/Deep-RL-Notes/blob/master/ddpg .tex

---

Algorithm 4.2 Q-learning: learn function $Q : \mathcal{X} \times \mathcal{A} \to \mathbb{R}$.

Require:

States $\mathcal{X} = \{1, \ldots, n_x\}$

Actions $\mathcal{A} = \{1, \ldots, n_a\}, \qquad A : \mathcal{X} \Rightarrow \mathcal{A}$

Reward function $R : \mathcal{X} \times \mathcal{A} \to \mathbb{R}$

Black-box (probabilistic) transition function $T : \mathcal{X} \times \mathcal{A} \to \mathcal{X}$

Learning rate $\alpha \in [0, 1]$, typically $\alpha = 0.1$

Discounting factor $\gamma \in [0, 1]$

1: **procedure** QLEARNING($\mathcal{X}, A, R, T, \alpha, \gamma$)

2:     Initialize $Q : \mathcal{X} \times \mathcal{A} \to \mathbb{R}$ arbitrarily

3:     **while** Q is not converged **do**

4:         Start in state $s \in \mathcal{X}$

5:         **while** s is not terminal **do**

6:             Calculate π according to Q and exploration strategy (e.g., $\pi(x) \leftarrow \arg\max_a Q(x, a)$)

7:             $a \leftarrow \pi(s)$

8:             $r \leftarrow R(s, a)$ ▷ Receive the reward

9:             $s' \leftarrow T(s, a)$ ▷ Receive the new state

10:            $Q(s', a) \leftarrow (1 - \alpha) \cdot Q(s, a) + \alpha \cdot (r + \gamma \cdot \max_{a'} Q(s', a'))$

11:            $s \leftarrow s'$

    return Q

Algorithm 4.3 Deep deterministic policy gradient (DDPG).

1: **while** true **do**

2:     Take some action a_i and observe (s_i, a_i, s_i', r_i) and add it to B

3:     Sample mini-batch $\{s_j, a_j, s_j', r_j\}$

4:     Compute $y_j = r_j + \gamma Q_{\phi'}(s_j', a_j')$ using target networks $Q_{\phi'}$ and $\mu_{\theta'}$

5:     $\phi \leftarrow \phi - \alpha \Sigma_j \frac{dQ_\phi}{d\phi}(s_j, a_j)(Q_\phi(s_j, a_j) - y_J)$

6:     $\theta \leftarrow \theta + \beta \sum_j \frac{d\mu}{d\theta}(s_j)\frac{dQ_\phi}{da}(s_j, a)$

7:     update ϕ' and θ'

---

waveform with a power spectrum (PS) containing a low-power notch to support spectrum sharing. To this end, DDPG supports a robust algorithm for optimal phases selection.

4.14 Application: Tracker's Parameter Optimization

In order to further exemplify the taxonomy proposed above, in this section an Off-Policy, offline, model-free, value-based DRL algorithm for continuous state-actions spaces is presented. This is utilized for solving a real-life use-case, namely a people-tracking task on short-range radar data.

Multitarget tracking with radars is a highly challenging problem due to detection artifacts, sensor noise, and interference sources. The traditional signal processing chain is, therefore, a complex combination of various algorithms with several tunable tracking-parameters. Usually, these are initially set by engineers and are independent of the scene tracked. For this reason, they are often nonoptimal and generate poorly performing tracking.

In this context, scene-adaptive radar processing refers to algorithms that can sense, understand, and learn information related to detected targets as well as the environment, and adapt its tracking-parameters to optimize the desired goal. In this use-case, we propose a DRL framework which guides the scene-adaptive choice of radar tracking-parameters toward an improved performance on multitarget tracking.

4.14.1 Motivation

In the modern semiconductor industry, radar sensors are gaining momentum. In fact, for common tasks such as people detection and object tracking, they allow for privacy, weather condition independence, and speed detection, as shown in [17, 18]. These factors, combined with the low-cost of the sensors, are the major causes for the market penetration of radar-based devices. In this context, customers steadily require improved features and capabilities. One of the main application fields of radar sensors is multitarget tracking (i.e., when multiple detection instances are generated for multiple targets), particularly in the domain of autonomous vehicles or human–machine interfaces, as described in [17, 19].

Multitarget tracking refers to the problem of maintaining the states of trajectories over time from unlabeled data sequences. Compared to single-target tracking, multitarget tracking has several challenges arising: detection errors, occlusions, and artifacts due to measurement noise, etc. For multitarget tracking, track filtering is usually achieved with a Kalman filter that allows exact inference using recursive equations estimating the state of the target under a dynamic model.

As the tracking performance highly depends on the underlying dynamic model, tracking-parameters (e.g., process angle variance, model acceleration) are usually set once by engineers to roughly fit the overall dynamics of the expected scenes. Nonetheless, as radar scenes are highly complicated due to factors like the sheering angle/orientation of the radar and arbitrary dynamics of the tracked target, the chosen tracking-parameters are often suboptimal for multiple scenes, resulting in poor tracking performance. To tackle this problem, approaches that consider an optimal combination of tracking-parameters for each scene (i.e., scene-adaptive) are required. As the parameters describe the underlying dynamic model in a continuous space, this leads to a continuous combinatorial optimization problem.

In the literature, several approximation approaches have been used to tackle the complexity of high-dimensional optimization problems. In particular, late contributions focused on the use of RL, such as in [15, 20–22].

Using RL, the reward signal directs the correct tracking behavior and guides the tuning of the tracking-parameters, for each of the scenes. Nevertheless, setting up such an RL framework for tracking-parameter optimization is not straightforward, in particular, if the following challenges are considered:

1. How to preprocess the radar signal so that it can be used as input to an RL framework for tracking-parameter optimization?
2. How to formalize a reward which encourages an optimal choice of the tracking-parameters in the RL framework?
3. How to improve the RL training by using additional tasks, which are not expensive in terms of labeling and equipment?

In this use-case, we address these questions. In particular, we first describe the preprocessing steps for adapting radar data to the RL framework. Afterward, we propose two different types of reward formulations, which adapt to the dynamic of the unscented Kalman-filter tracking. These guide the system toward the choice of the optimal scene-adaptive tracking parameters. Finally, we propose auxiliary tasks, which involve little manual effort but significantly contribute to the performance of the RL framework. Results show that, utilizing the proposed approaches, we can get 226% and 93% closer to the optimal return on the two reward formulations, when compared to the baseline tracking-parameters. Furthermore, the miss-prediction rate is decreased by a factor of 6×.

The remainder of this use-case is structured as follows: in Section 4.14.2, we present the background related to signal processing and RL; in Section 4.14.3, we describe the proposed methods; while in Section 4.14.4, we evaluate them. Finally, Section 4.14.5 concludes the section.

4.14.2 Background

In this section, we describe the background and motivation for our contribution. To this end, we first describe the traditional signal-processing pipeline for multitarget tracking, and then we explain concepts of RL that extend it toward scene-adaptive tracking parameters. Finally, we give an overview of RL on related work of multitarget tracking.

4.14.2.1 Traditional Signal Processing Tracking for Radar

Human target tracking with radars as a research topic has now been studied for multiple decades. Typical approaches for frequency modulated continuous wave (FMCW) radar sensors include target detection, range, angle, and velocity estimation, target association, and track filtering. The authors in [23] describe a tracking method based on fusing camera and radar data. They first extract range and Doppler information from the radar signal via a 2D fast Fourier transform (FFT), then identify potential targets with a constant false alarm rate (CFAR) detector and estimate the associated angles before combining these results with the camera detections for track filtering. For the purpose of adaptive driver assistance systems, the authors in [24] use a similar radar processing chain. They add another clustering block after the CFAR detection to reduce the false alarms and to get the center locations. Instead of tracking, they then use this processed information as an input for a neural network for target detection and classification. In [25], the authors add a coherent pulse integration block after the range FFT and estimate the angle and velocity only after peak detection with adaptive thresholding to save processing power. In the same vein, they then use an alpha–beta filter as a light-weight tracking solution. We use a similar radar processing chain for target tracking as shown in Figure 4.3, at each timestep, we receive one 2D radar data frame from each receiving antenna, containing 128 fast time samples and 64 chirps each. Within these 2D data frames, we first need to detect potential target returns and extract their range angle and velocity values, before feeding those measurements into our tracker.

4.14.2.2 Potential Target Detection and Parameter Estimation

We transform the 2D time-domain data to range-Doppler images (RDI) via a 2D FFT and then remove any completely static targets with a 2D moving target indication (MTI) filter. To get rid of spectral leakage from peaks into adjutant range- or Doppler bins, we use a Blackman window for the fast-time data and a Chebyshev window for the slow-time data before doing the respective FFTs, as shown in [26]. The MTI filter is an exponential moving average low-pass filter, whose output,

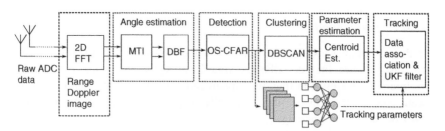

Figure 4.3 Traditional and proposed processing pipeline for radar target tracking.

containing mainly completely static targets, is subtracted from the current frame. Afterward, we use digital beamforming (DBF) to transform the RDIs from the two receiving antennas to a range angle image (RAI). The ordered statistics constant false alarm rate (OS-CFAR detector block, as described in [27], then detects potential target returns based on the estimated local noise level in the RAI. These detections are then clustered based on their distances using the density-based spatial clustering of applications with noise (DBSCAN) algorithm [28]. The inputs to the tracker, an unscented Kalman filter (UKF), as described in [29], in this case, are then the estimated parameters, namely, range, azimuth angle, and velocity, of each cluster centroid.

4.14.2.3 Kalman Filter

Kalman filtering [30] consists of two main steps, the predict-step and the update-step. In the predict-step, the state and its uncertainty after the measurement from the next frame is predicted from the current state and its uncertainty, a potential correction based on known influences, a state transition matrix, and another covariance matrix modeling the uncertainty in the state transition matrix. In the update step, the predicted state and its uncertainty are used along with the new measurement and its measurement uncertainty, to update the current state and its uncertainty. The Kalman filter assumes that the mappings $x_{k+1} = F(x_k, v_k)$ and $z_{k+1} = H(x_k, n_k)$ describe a linear relation between the current state x_k, some process noise v_k and the next state x_{k+1}, and the current state, some measurement noise n_k and the measurement z_{k+1}, respectively.

4.14.2.4 Unscented Kalman Filter

If the mapping functions $F(x_k, v_k)$ or $H(x_k, n_k)$ are nonlinear, the issue lies in calculating the nonlinear transformations of the predicted Gaussians. To solve this problem, the UKF uses an approximation for the probability mapping via the unscented transform [31], which is then used in the processing chain to realize the overall process and measurement model. The basic idea of the unscented transform is to apply the nonlinear transform to sampled points from the original Gaussian and to use the transformed points to find a new Gaussian to approximate the transformed distribution with. Specifically, one uses some carefully chosen sigma points $\mathcal{X}$ as follows:

$$\mathcal{X}^{[0]} = \mu \tag{4.9}$$

$$\mathcal{X}^{[i]} = \mu + \sqrt{\frac{n}{1 - W_0}} \Sigma_i^{0.5}, \quad i = 1, 2, \ldots, n \tag{4.10}$$

$$\mathcal{X}^{[i]} = \mu - \sqrt{\frac{n}{1 - W_0}} \Sigma_i^{0.5}, \quad i = n+1, n+2, \ldots, 2n \tag{4.11}$$

where $W_i = \frac{1-W_0}{2n}$ and $\Sigma_i^{0.5}$ are the ith column of $\Sigma^{0.5}$, which is the Cholesky decomposition of the covariance matrix Σ, μ, and Σ are the mean and covariance of the original Gaussian, and n is the dimension of the state [32].

In our radar processing chain shown in Figure 4.3, the state vector x_s is given by $x_s = [x, y, v, \phi, \dot{\phi}]$, where x and y are the Cartesian coordinates, v the radial velocity, ϕ the azimuth angle, and $\dot{\phi}$ the azimuth acceleration of the tracked target. With each measurement consisting of the range $\rho = \sqrt{x^2 + y^2}$, angle ϕ and radial velocity v, nonlinearities appear in both $F(x_k, v_k)$ and $H(x_k, n_k)$. The nonlinear state transition, as seen in [33], is given by

$$
\begin{pmatrix} x \\ y \\ v \\ \phi \\ \dot{\phi} \end{pmatrix} = \begin{pmatrix} x + \frac{2v}{\phi} \sin\left(\frac{\phi T}{2}\right) \cos\left(\phi + \frac{\phi T}{2}\right) \\ y + \frac{2v}{\phi} \sin\left(\frac{\phi T}{2}\right) \sin\left(\phi + \frac{\phi T}{2}\right) \\ v \\ \phi + \dot{\phi} T \\ \dot{\phi} \end{pmatrix}.
\tag{4.12}
$$

One challenging part in making Kalman filters work is setting its tracking-parameters sensibly. For the human target-tracking task at hand specifically, we use 14 adjustable tracking-parameters in the tracker implementation. These consist of five process noise parameters, six measurement noise parameters, two parameters for creating and deleting tracks, and a Mahalanobis distance-based gating parameter to decide whether a measurement should be associated with any track. The measurement noise variances along range, Doppler, angle, and the range-angle, range-Doppler and angle-Doppler covariances are generally connected to the radar specifications and therefore easier to set. The process noise parameters, however, may require more careful tuning. One way of tuning these parameters is by using the normalized innovation squared (NIS) metric, defined as follows:

$$
\text{NIS}_{k+1} = \epsilon_{k+1} \Omega_{k+1}^{-1} \epsilon_{k+1}^H
\tag{4.13}
$$

NIS_{k+1} is χ^2- distributed with m degrees of freedom, where m is the number of measurement variables, ϵ_{k+1} is the innovation gained with each new measurement, and Ω_{k+1}^{-1} is the inverse of its covariance matrix [26, 34]. Figure 4.4 shows the NIS for two different tracking-parameter settings. In Figure 4.4a, the NIS is visibly above the horizontal line visualizing the 95% confidence interval. This indicates that the new measurement values z_{n+1} are mostly far off the predicted values $\hat{z}_{n+1|n}$, implying an underestimation of the noise values. Figure 4.4b shows the opposite scenario, where the NIS mostly falls below the 5% confidence region, indicating that the noise variances are set too high as sketched by $\hat{z}_{n+1|n}$ lying close to z_{n+1}. In our case, the tracking-parameters are manually chosen by an engineer, starting from a set of heuristic parameters based on the radar system parameters and expected use-cases, followed by a guided grid search using the NIS score

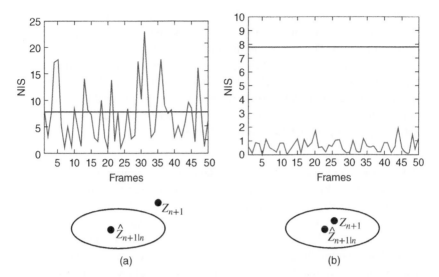

Figure 4.4 (a) NIS is higher than 95% confidence score and (b) NIS is lower than 5% confidence score.

and ground-truth position data as a performance metric on a set of diverse indoor recordings (mainly office spaces and apartment rooms).

Since depending on the scene and deployed environment and tracked targets, the optimal parameters for the tracker could vary drastically and are always challenging to find. In fact, there may also not be one set of optimal tracking-parameters.

This problem has been often tackled in the scientific literature, e.g., from these contributions [35, 36]. In the first paper [35], the authors design a system which takes into account the trade-off between the tracking precision and the loss-of-lock condition. To this end, they elaborate a sequence of linear predictors for determining the target motion, with a fall-back mechanism. This mechanism, once the linear models fail (for example because of the saccadic motion of the target), automatically gives up the precision in order to avoid loss-of-lock. In this following work [36], different methods for tracking-parameters optimization of visual Trackers are proposed and compared. While in both papers sophisticated approaches for retrieving tracking-parameters are provided, none of the methods presented elaborate long-term strategies based on the state transition, action, and cumulative future reward. This means, the choice of the tracking-parameters depends in their case, only on the immediate outcome of the tracking precision, and does not exploit exploration strategies for selecting tracking-parameters which may start as suboptimal. Additionally, in this paper [35], the focus lays on selecting a series of linear predictors for tracking the target motion, thus limiting the motion prediction of the target. Finally, in this contribution [36], methods presented start with a general model of the tracking-parameter optimization, determined a priori. This limits the flexibility of the tracking-parameter estimation model. In order to address these drawbacks, RL is introduced as an effective and yet scalable approach to tackle the search for optimal tracking-parameters. To this end, in Section 4.14.2.5, we analyze how DRL can be used for continuous optimization. After it, in Section 4.14.2.6, we describe how this improved the search for tracking-parameters for different sensors.

4.14.2.5 Deep Reinforcement Learning for Continuous Optimization

While historically DRL methods have been focusing of discrete action spaces, both On-Policy and Off-Policy RL methods have been proven effective for continuous optimization. While On-Policy methods attempt to evaluate or improve the policy that is used to make decisions, Off-Policy methods evaluate or improve a policy different from that used to generate the data. As a consequence, Off-Policy methods are often used in real-life scenarios, as they can reutilize methods collected from a different policy. To this end, the guarantee of a major sample efficiency and allowance to learn from experts demonstrations lead to a major adoption in real-life scenarios.

Within the Off-Policy methods used for continuous action spaces, DDPG [15] and normalize advantage function (NAF) [21] are the most frequently adopted. They deal with continuous action space by learning the maximum action for a given state. While DDPG is an actor-critics algorithm which concurrently learns a Q-function and a policy, NAF utilizes only continuous Q-learning paradigm, thus simplifying the optimization.

However, since in these methods a deterministic policy is learned, exploration is hard to obtain. Yet one of the main issues when applying DRL in real-world problems is how to design the reward function when this one is sparse, episodic (given only at the end of the episode) or even not explicit at all. inverse reinforcement learning (IRL) [37] tries to address this issue by learning a reward function (instead of a policy) that best explains a desired behavior contained in a (large) collection of expert demonstrations. Although IRL methods might not be suitable when expert demonstrations are scarce, it brings an interesting result of how to deal with suboptimality in the expert demonstrations by the principle of maximum entropy: keeping the policy's entropy high while maximizing the return results in a better exploration which produces a more robust policy that is easier to transfer. Similar results can be found in [38].

So far, several RL approaches have been applied to tasks related to radar multitarget tracking. These have been summarized in Section 4.14.2.6.

4.14.2.6 Multitarget Tracking via Reinforcement Learning

Methods of multitarget tracking via RL have been explored in the computer vision (CV) field [20]. Here, a scene-adaptive tracker, trained via DRL, is proposed. While this work paves the way to further exploration of DRL for tracking-parameters optimization, it presents some challenges toward its implementation on radar trackers. In fact, radar input data require a more complex preprocessing than visual data. Additionally, the amount of tracking-parameters presented in that contribution is considerably lower if compared to the settings of an UKF, which in turn increase the problem space. Moreover, the proposed reward formulation in the paper [20] is specifically related to the overlapping of the prediction position to the ground truth and does not encourage tracking-parameters exploration in the DRL problem. Finally, the solution does not use auxiliary tasks, which could be potentially helpful for a better convergence of the DRL algorithm. On the other hand, multitarget tracking via RL has been explored also within the radar signal processing community. In [39], a long-short-term memory network (LSTM) is proposed to learn the measurement-to-track association probability to solve the non-deterministic polynomial-time (NP) hardness combinatorial association problem in multitarget tracking systems. Additionally, in the systems, tracking highly maneuvering targets requires executing several filters in parallel, either in the input estimation approach or the interacting multiple model approach. However, such approaches are not conducive for resource-constrained implementations, thus in [40] a hybrid neural network approach is proposed to improve tracking accuracy. Cognitive radars (CRs) can adapt their operations based on their environment and accumulate knowledge from their interactions with the environment. To this end, RL has been proposed in several such proposed solutions such as in [41–43].

In [41], authors propose cognitive imaging using RL by adapting the transmitted waveform. In [43], authors propose to solve the NP-hard problem of radar resource management using modified Monte Carlo tree search, which is further guided by a RL framework. In [44], authors propose a reinforcement learning-based DQN to solve the target assignment problem in a phased-array radar network. In [45], the problem of multitarget detection in massive MIMO cognitive radar is considered, and a cognitive multitarget detection in the presence of unknown disturbances using RL is proposed. Finally, in [46], the radar-communication coexistence problem is modeled by an MDP and solved using RL. Although, as described above, several RL approaches have been developed to improve radar multitarget tracking, no contributions yet optimize the tracking-parameters (e.g., process angle variance, process noise) in a scene-adaptive fashion. In the next section, we introduce the proposed approach, which enables the choice of scene-adaptive tracking-parameters.

4.14.3 Approach

In this section, we describe the proposed approach toward scene-adaptive tracking-parameters. To this end, (1) we first analyze the preprocessing steps introduced for radar data; (2) then we discuss the formulation of the RL problem and the training phases within the proposed RL framework; (3) and finally, we describe the suggested networks pretraining through an auxiliary task.

4.14.3.1 Data Processing

As a first step of the data preprocessing, we use the pipeline presented in Figure 4.3 for obtaining an easy-to-process and informative input for the tracking-parameters learning process. As an input to the tracking-parameters RL framework, we use the RAI after OS-CFAR detection. These masked RAIs contain information about the environment, human, and nonhuman targets, as well as possible ghost targets. To allow also the extraction of track-related information and to become less susceptible to random perturbations and occlusions, the input to the RL framework consists of the data from eight successive frames. This is consistently used during the training phases of the RL framework.

4.14.3.2 Training Phases

In this section, we describe the training phases for learning optimal scene-adaptive tracking-parameters. In our settings, the action space corresponds to the set of tracking-parameters, which are continuous values. This fact alone justifies the use of DL techniques in our RL method. Since the action space is continuous, complex approximation as well as generalization capabilities are needed in the method. DL allows for such continuous space approximation and generalization, and hence, justifies the use of deep RL methods. The state consists of the number of tracks and the positions for the ground-truth and the number of predicted tracks together with their corresponding information about predicted positions. Actions change the internal state of the tracker predictor. The choice of RL framework is justified since we aim at optimizing the set of tracking-parameters at every timestep for the whole sequence, taking into account long-term dependencies. The optimal set of tracking-parameters can be different depending on if the future is considered or not. In the RL settings, actions are chosen to maximize the cumulative sum of rewards in a sequence, instead of maximizing every timestep alone. The reward function is proposed below. As the optimal tracking-parameters are chosen within continuous sets, we follow the NAF approach to learn a policy that selects the best tracking-parameters at every scene. To this end, we implement three neural networks: *actor network* μ, from where we obtain the tracking-parameters; *lower triangular network P*, which recreates the state representation through reconstructing a positive-definite matrix; and the *value network V*, which approximates the value function. Additionally, a target network is used to stabilize the Bellman error when learning the *Value Network*.

$$Q(s, a; \theta^Q) = A(s, a; \theta^A) + V(s; \theta^V) \tag{4.14}$$

$$A(s, a; \theta^A) = -\frac{1}{2}(a - \mu(s; \theta^\mu))^T P(s; \theta^P)(a - \mu(s; \theta^\mu)) \tag{4.15}$$

This approach allows us to subdivide the training process into three phases, involving different neural networks at a time, combining imitation learning and RL in a coordinate fashion. By splitting the training process into phases, it is possible to monitor and encourage the improvement of each single network. Similar to [20], Phase I applies imitation learning on a set of demonstrations to initialize the *actor network*. Demonstrations contain a sequence of scenes and their fixed tracking-parameters, manually tuned by human experts. After this phase, the *actor network* learns to select the tuned tracking-parameters for a given scene.

In Phase II, the *actor network* is fixed, and the *value network* and *lower triangular network* are trained to minimize the Bellman squared error. The *value network* is copied and fixed to serve as target network. This target network is updated during learning using a Polyak average of the current *value network* entropy.

Finally, in Phase III, all the three networks are trained jointly via RL in a synchronized fashion. The outcome of this phase is an improved *actor network* able to output the optimized scene-adaptive tracking-parameters.

4.14.3.3 Reward Formulation and Model-Based Variance

Designing a reward function that efficiently leads to the desired behavior is one of the most critical components when applying RL to real-world problems. In our application, the environment provides at every time step the number of tracks $N_t \in \mathbb{N}$ and positions $P_t \in \mathbb{R}^2$ for the current scene provided by the cameras. Ideally, the actor should select the optimal tracking-parameters that result in the correct number of predicted tracks $\hat{N}_t$ and their correct position prediction $\hat{P}_t$ along the whole sequence. Therefore, the reward function should have two components: one to measure how accurate the system is at predicting the number of tracks (ρ); and a second term to measure how far the positions are with respect to the ground truth (d). Since the optimal behavior would lead to zero, and our policy is optimized to maximize the return, we change the sign of the reward.

$$-R_t = \rho(\hat{N}_t, N_t) + d(\hat{P}_t, P_t) \tag{4.16}$$

The first term ρ is implemented as the relative error of the predicted and true number of tracks. For the second term, d, we propose two variants. The first variant, (d_1), implements d as the sum of Euclidean distances between the true and the predicted track positions. For the predicted tracks, only the predicted mean of the Gaussians is considered. The procedure is described in Algorithm 4.1. The second

Algorithm 4.4 Ordered minimum assignment.

Input: $X \in R^{Mx2}$: predicted track means
1:      $Y \in R^{Nx2}$: position labels
Output: distvals: list of Euclidean distances between assigned tracks
2: *Initialization*:
3: C = cost_matrix, where c_{ij} is the Euclidean distance between the predicted track mean $x[i, :]$, and the position label $y[j, :]$, $C \in R^{MxN}$
4: $K = \max(C)$
5: distvals = [ ]
6: **for** $i = 0$ to $(min(M, N) - 1)$ **do**
7:      row_idx, col_idx = argmin(C)
8:      $C[row\_idx, :] = K$
9:      $C[:, col\_idx] = K$
10:     append $C(row\_idx, col\_idx)$ to distvals
    return distvals

variant, (d_2), also takes into account the covariance matrix Σ provided by the UKF.

$$d_2(\hat{P}_t, P_t) = \frac{1}{M} \sum_{k=0}^{M} 1 - p_k(P_k) \tag{4.17}$$

where $M = \min(\hat{N}, N)$, $p_k \sim N(\hat{\mu}_k, \hat{\Sigma}_k)$ and P_k is the ground truth position for the kth track. To maintain stability, we clip the $p_k(\cdot)$ value to 1. This variant computes how likely it is to miss the ground truth position by evaluating the true position P_k on the multivariate Gaussian distribution given by its mean vector $\hat{\mu}_k$ and its covariance matrix $\hat{\Sigma}_k$. This variant is meant to encourage exploration. Since the NAF approach results in a deterministic policy, exploration is hard to obtain with standard procedures for value-based methods (i.e., ϵ-greedy). Inspired by the results of maximum entropy IRL [47], d_2 injects some small reward when the tracking-parameters result in a high variance Kalman filter output. If the mean vectors of the predictions are off with regard to the true positions, then having higher variances in the predicted track distributions will result in a higher reward as seen in Figure 4.5. The intuition behind this additional reward is to achieve better robustness by promoting exploration when the current tracking-parameters lead to suboptimal tracking. As a last remark, it is important to note that the optimum of the two reward variants is 0.

4.14.3.4 Auxiliary Tasks and Pre-training for Transfer Learning

Auxiliary tasks can often improve the performance of ML models. Auxiliary tasks are particularly important when labeled data are limited. As formulated in Eq. (4.16), the reward depends on some distance between predictions and the true target labels. For this reason, the target position should be provided as a label.

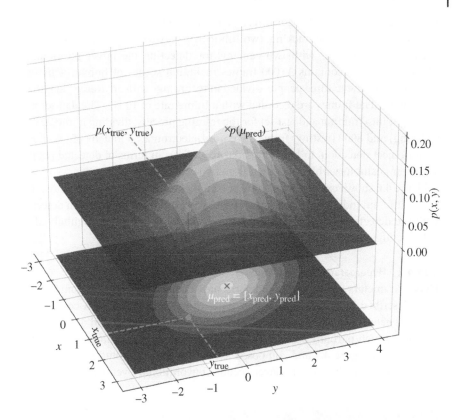

Figure 4.5 Multivariate Gaussian representing the predicted track distribution. The reward for the predicted target also depends on its associated variance by taking into account the value of the Gaussian at the true target position $p(x_{true}, y_{true})$.

However, the position labeling requires no small amount of manual effort, and typically entails depth-cameras and computationally expensive preprocessing. To reduce computational costs, we pretrain our model on radar data to predict the number of people in each scene and additionally classify whether the target is static or dynamic. Labels for this task are easy to obtain using only the radar sensors and therefore, large amounts of data can be exploited. After training, the model learns useful information that can be transferred to the main task.

4.14.4 Experimental

4.14.4.1 Implementation Settings and Dataset

In the implementation, we used TensorFlow™ - GPU v2.4.0 with CUDA® Toolkit v11.1.0 and cuDNN v8.0.5. As processing unit, we used Nvidia® Tesla® P40 GPU, Intel® Core i7-8700K CPU and DIMM 16GB DDR4-3000 module of RAM.

For implementing the RL algorithms, we used the TF-Agents framework, v0.7.1. In order to perform the tracking, two Infineon's XENSIV™ 60 GHz sensors have been utilized together with two cameras for detecting the target positioning. The dataset used contains 30 000 frames including scenes of human activities performed by one to three people, divided in recordings with an average length of 350 frames. The frames are recorded with a frame rate of 10 Hz. The dataset has been split into training and test set for the training process, dividing it into 22 000 (training) and 8000 (testing) frames. We use Detectron 2,[4] a toolbox of Facebook AI for object detection, to label the radar recordings in terms of detected people and their positions with information from two cameras.

Figure 4.6 shows examples for a one person recording and a two person recording in different indoor environments. The joint positions for each human detected with the detectron framework and used for position labeling are highlighted in both pictures. The radar position is indicated with circles.

4.14.4.2 Hyperparameters and Training

Hyperparameters for the neural networks have been retrieved by random search.

As described in Section 4.14.3, four types of networks have been considered, an *actor network*, a *lower-triangular network*, a *value network*, and a *pre-trained network*, which partially share the same architecture. The common structure is visualized in Figure 4.7 and composed as follows: first convolutional layer (128 5×5 kernels), first batch-norm layer, second convolutional layer (64 3×3 kernels),

Figure 4.6 Indoor recording environment example pictures, taken with one of the cameras used for labeling purposes.

---

4 https://github.com/facebookresearch/detectron2

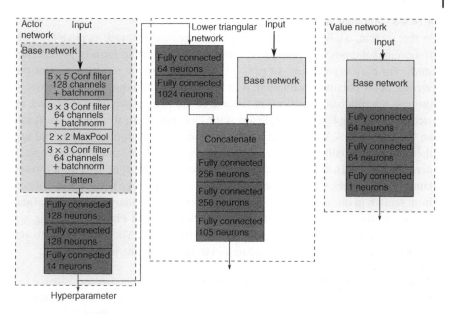

Figure 4.7 Respective network architectures of the actor network, lower triangular network, and the value network used for the NAF learning.

second batch-norm layer, max-pooling layer (2 × 2 kernel with stride 2), third convolutional layer (64 3 × 3 kernels), third batch-norm layer, first dense layer (256 neurons), second dense layer (128 neurons), third dense layer (64 neurons), output. Further, additional layers are implemented onto the common structure to form the *actor network*, *lower-triangular network*, and the *value network*, as shown in Figure 4.7.

For the *actor network*, *value network*, and *pretrained network*, the input is composed by eight successive frames of radar-preprocessed data (32 × 29 × 16). For the *lower-triangular network*, the tracking-parameters vector (1 × 14) sourced from the *actor network* is first preprocessed by two dense layers, and then concatenated to the output of the convolutional layers. For the three networks, we use Adam optimizer, described in [48]. Learning rates are $1e - 5$, $1e - 6$, and $1e - 5$ for the auxiliary task initialization, for Phases I, II, and III, respectively. For the plain initialization, $1e - 3$, $1e - 4$, and $1e - 3$. As specified in Section 4.14.3.2, three networks are then utilized for the three phases of the training process, while the *pretrained network* has been used for learning common features from an auxiliary task. The training phase of the *pretrained network* has 1k training steps. Regarding the successive training phases, we perform 10k steps in training Phase I, 200k steps in training Phase II, and 24k steps in training Phase III. In order to benchmark the propositions of this contribution in terms of tracking, in Section 4.14.4.3, we conduct experiments in the form of an ablation study.

4.14.4.3 Ablation Study and Results

In this section, we benchmark the impact of the auxiliary task as model initialization and the newly proposed reward formulation in the form of an ablation study. For each experiment, 10 runs are executed, and the results represent the average of the obtained evaluation. As a first step, we train the *pretrained network* through a multiclass classification task on the number of people in the scene, as well as predicting if those people are static or dynamic. In this phase, we reach an F1-score of 92%. We then detach the three initial convolutional layers and use them for the initialization of *actor network*, *value network*, and *lower-triangular network*. After this, we first subdivide the experiments in d_1 or d_2 reward formulation. Successively, within these categories, we use the three networks initialized from scratch in one case, or initialized by performing auxiliary tasks in the other. In each of the single experiments, Phase I, Phase II, and Phase III training are performed, and then the model is evaluated on the test set. The best *actor networks* models for d_1 and d_2 are then compared in terms of tracking prediction.

4.14.4.4 Euclidean (d_1) and Variance-Aware Reward (d_2)

In Figure 4.8a, results in terms of return on test data are shown for *actor network* models trained with an Euclidean reward. The return obtained by utilizing a plain initialization mode is -1.18, while when using a pretrained architecture on auxiliary tasks, we obtain up to -0.92. The results visualized represent the test set outcomes, using the plain or pretrained network (i.e., using the initialization from the auxiliary-task training).

As shown, by extracting the relevant features, the pretrained model is able to increase its convergence limit and obtain an higher reward with regard to the plain initialized model. For the d_1 reward variant, the outcome of manually tuned tracking-parameters is -2.08 (i.e., baseline).

In Figure 4.8b, results are shown for the models trained with variance-aware reward. The return obtained by utilizing a plain initialization model is -0.68, while when using a pretrained architecture on the auxiliary task, we obtain up to -0.58. As visualized in Figure 4.8b, the pretrained model is able to outperform the plain model.

For this type of reward, the reward outcome of the baseline is -1.06.

As it is possible to observe from Figure 4.8a and b, the convergence of the pre-trained model is obtained only at later steps of the training process. For this reason, in order to ensure the completion of the learning process, the model requires the whole amount of training steps in Phase III.

4.14.4.5 Tracking Prediction

In order to benchmark the tracking prediction obtained by the proposed reward formulations, and the baseline tracking-parameters tuned manually, we compare their error on a set of evaluation recordings. To this end, we utilize the

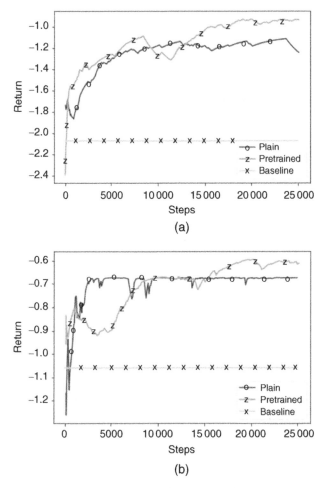

Figure 4.8 Reward development on the test set over the steps of the Phase III training for plain vs. pretrained vs. baseline using variant d_1 (a) and d_2 (b).

best-performing models obtained by using reward variants d_1 and d_2. In this benchmark, the error has been divided in terms of false alarms/miss-detection as well as distance to ground truth.

Figures 4.9 and 4.10 show scene tracking performed using the model trained with reward variant d_2, on two test scenes with around 650 frames (Figure 4.9) and 350 frames (Figure 4.10). The scene tracking shown in Figure 4.9 has the highest tracking accuracy and lowest miss-prediction rate within the evaluation recordings for the proposed reward variant. Figure 4.10 instead shows the worst-performing scene tracking for the proposed reward variant.

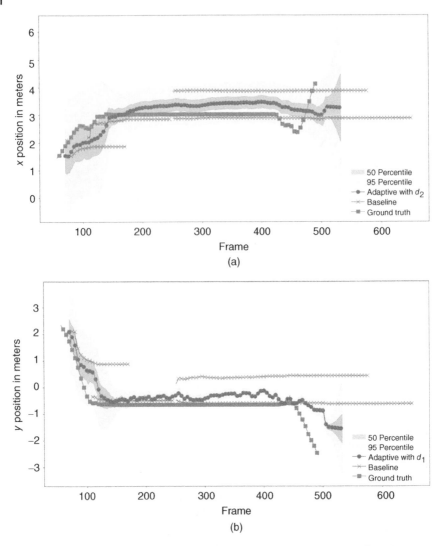

Figure 4.9 Best-performing scene position predictions, pretrained (d_2) vs. Baseline on (a) x and (b) y coordinates for a one-target scene.

In both cases, the proposed method visibly outperforms the baseline (a comparable distance from the ground truth has been obtained by using the variant d_1). In those particular test, scenes the baseline tracking detects the true target trajectory for the majority of frames, but fails to reject false targets and fails to follow sudden changes in the true target position. For both Figures 4.9 and 4.10, the 50. and the 95. percentile for the multivariate Gaussian track prediction output of the Kalman

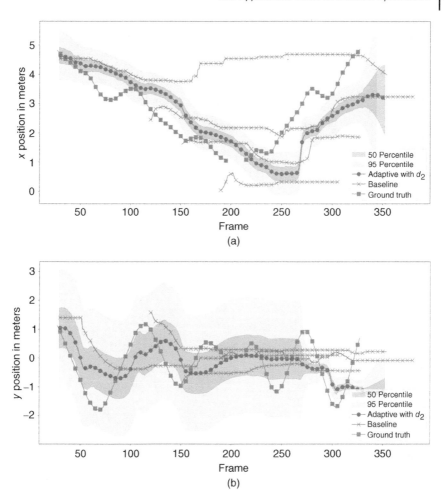

Figure 4.10 Worst-performing scene position predictions, pretrained (d_2) vs. baseline on (a) x and (b) y coordinates for a one-target scene.

filter are shown for the x-position in (a), and for the y-position in (b). By using d_2, the tracker tends to increase the uncertainty of the predictions, if the true target is further away from the prediction's mean. Accordingly, in both figures, the variance is larger when the target enters and exits the field of view, and when the true target is out of the field of view, but the tracker mistakenly still detects it. The variance is visibly smaller when the predictions are closer to the true target between frames 100 and 400 in Figure 4.9, and between frames 270 and 350 in Figure 4.10. Figure 4.11 shows another comparison of the proposed method to the baseline of an office scene, where a person walk up to a chair, sits down for some

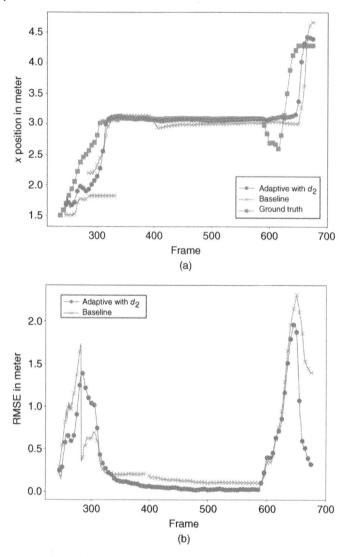

Figure 4.11 Comparison of (a) *x* position prediction for pretrained (d_2) vs. baseline and of the (b) RMSE of the closest prediction to the ground truth.

time, and walks away again. Figure 4.11a shows the *x*-position predictions of both chains compared to the ground truth (dotted line), while Figure 4.11b shows the corresponding root-mean-square error (RMSE) values between the closes predicted target to the ground truth. In this particular scene, both processing chains manage to track the target with a short delay on accelerations. While the adaptive chain detects the one target throughout the scene, the baseline falsely detects two targets for about 40 frames.

On the whole test frames of the evaluation recordings, the average Euclidean distance obtained for the best model using the reward variant d_1 is 0.9 m, with an average false alarms/miss-detection error of 0.25. On the other hand, for the best model using reward variant d_2, the average Euclidean distance is 1.0 m, while the false alarms/miss-detection error is 0.25. For the baseline, the values for the Euclidean errors and false alarms/miss-detections are 0.93 m and 1.5, respectively.

4.14.4.6 Statistical Relevance of the Variance-Aware Reward

As a last experiment, we determine the statistical relevance of the proposed reward variant d_2. To this end, we select the best performing, pretrained models using reward variants d_1 and d_2, and we evaluate them using the reward variant d_2 on an subsequent interval of 4000 frames, randomly selected out of the test set. Our objective is determining if training with reward variant d_2 would lead to a significantly different distribution of returns with regard to training with reward variant d_1. For reward variant d_2, we obtained a mean evaluation return of -0.50 and variance 0.02. For the model trained with reward variant d_1, the mean return was -0.69 and the variance 0.05. Since we cannot assume that the returns are normally distributed, we perform the Wilcoxon signed-rank test to assess significance to our results. In our research, this test is used for assessing if reward variants d_1 and d_2 produce policies with the same performance. The outcome of this test shows a $p = 0.00$, rejecting the null hypothesis and showing that the policy trained with reward variant d_2 are significantly better than the policy trained with reward variant d_1.

4.14.5 Outcomes of the Proposed Approach

In this section, we present a method for tracking-parameter optimization of radar sensors, using RL techniques for continuous optimization. To this end,

1. We propose an ad hoc data processing pipeline, in order to obtain input data for the RL framework.
2. We introduce two reward variant formulations. The first reward variant uses Euclidean distances, while the second takes the variance of the Kalman filter prediction into account.
3. We propose auxiliary tasks on the RL training, which does not use sophisticated positional labels.

Experiments show that our approach achieves a significant improvement for the first reward variant (i.e., -0.92, 226% closer to the optimum), and for the second reward variant (i.e., -0.58, 93% closer to the optimum), with regards to the manually tuned baseline parameters. Furthermore, both variants reduce the false alarms/missed detection by a factor of 6×.

These outcomes substantially increase the tracking performance. In turn, this leads to more efficient real-life solutions in the field as autonomous-driving, automated air-conditioning/heating regulation, personalized sound systems, and video sources.

4.15 Conclusion

In this chapter, we presented an overview of (deep) reinforcement learning. To this end, we first reviewed the general concept of RL. Afterward, we realized a taxonomy of RL algorithms, where different categories have been established:

- On-Policy/Off-Policy,
- Model-based/Model-free,
- Value-based/Policy-based,
- Online/Offline,
- Discrete Actions/Continuous Actions.

Finally, we describe a use-case where an Off-Policy, offline, model-free, value-based DRL algorithm for continuous state-actions spaces has been employed for target-tracking with radars. To this end, the proposed methods are able to overcome the state-of-the-art on short-range radar tracking and be 226% closer to the optimal tracking performance.

4.16 Questions to the Reader

- Which type of RL algorithms are more sample-efficient: Off-Policy or On-Policy?
- How does the size of a replay buffer affect the computational complexity of an RL algorithm?
- What are some problems where you would apply model-based RL algorithms?
- For a MDP-based problem, where the state/action space is extremely large, would you prefer to apply a value-based or policy-based algorithm?
- Imagine you are implementing an RL algorithm in an embedded radar sensor: what are the limiting factors? Which type of algorithm would help you overcoming them?
- Given what you read in this chapter, can RL in practice be applied, with good results, also to non-MDP problems? Explain why.
- You have to solve a specific problem and you decide to use RL. How would you design your reward function? What are the factors you would consider mostly, while designing it?
- How can a normalized advantage function help in the stability and convergence of an RL algorithm?

References

1 Sutton, R.S. and Barto, A.G. (2014). *Reinforcement Learning: An Introduction.* MIT Press.

2 Vinyals, O., Babuschkin, I., Czarnecki, W.M. et al. (2019). Grandmaster level in StarCraft II using multi-agent reinforcement learning. *Nature* 575 (7782): 350–354.

3 Mnih, V., Kavukcuoglu, K., Silver, D. et al. (2013). Playing Atari with Deep Reinforcement Learning.

4 van Hasselt, H. (2010). Double Q-learning. *Advances in Neural Information Processing Systems 23 (NeurIPS).*

5 Wang, Z., Schaul, T., Hessel, M. et al. (2016). Dueling network architectures for deep reinforcement learning. In: *Proceedings of the 33rd International Conference on Machine Learning (ICML),* ser. Proceedings of Machine Learning Research (ed. M.F. Balcan and K.Q. Weinberger). JMLR.org.

6 Billard, A., Calinon, S., Dillmann, R., and Schaal, S. (2008). Robot programming by demonstration. In: *Springer Handbook of Robotics* (ed. B. Siciliano and O. Khatib), 1371–1394. Springer.

7 Goodfellow, I., Bengio, Y., and Courville, A. (2016). *Deep Learning.* MIT Press.

8 Arulkumaran, K., Deisenroth, M.P., Brundage, M., and Bharath, A.A. (2017). Deep reinforcement learning: a brief survey. *IEEE Signal Processing Magazine* 34 (6): 26–38.

9 Rummery, G.A. and Niranjan, M. (1994). On-Line Q-Learning Using Connectionist Systems. *Tech. Rep. 166.* Cambridge University Engineering Department.

10 Wang, L., Fortunati, S., Greco, M.S., and Gini, F. (2018). Reinforcement learning-based waveform optimization for MIMO multi-target detection. *2018 52nd Asilomar Conference on Signals, Systems, and Computers.* IEEE, pp. 1329–1333.

11 Mnih, V., Kavukcuoglu, K., Silver, D. et al. (2013). Playing Atari with Deep Reinforcement Learning. *arXiv preprint arXiv:1312.5602.*

12 Thornton, C.E., Kozy, M.A., Buehrer, R.M. et al. (2020). Deep reinforcement learning control for radar detection and tracking in congested spectral environments. *IEEE Transactions on Cognitive Communications and Networking* 6 (4): 1335–1349.

13 Schulman, J., Levine, S., Abbeel, P. et al. (2015). Trust region policy optimization. *International Conference on Machine Learning.* PMLR, pp. 1889–1897.

14 Li, K., Jiu, B., Liu, H., and Pu, W. (2021). Robust antijamming strategy design for frequency-agile radar against main lobe jamming. *Remote Sensing* 13 (15): 3043.

15 Lillicrap, T., Hunt, J.J., Pritzel, A. et al. (2016). Continuous control with deep reinforcement learning. *ICLR* 2016.

16 Smith, G.E. and Reininger, T.J. (2021). Reinforcement learning for waveform design. *2021 IEEE Radar Conference (RadarConf21)*. IEEE, pp. 1–6.

17 Zou, Y., Liu, W., Wu, K., and Ni, L.M. (2017). Wi-Fi radar: recognizing human behavior with commodity Wi-Fi. *IEEE Communications Magazine* 55 (10): 105–111.

18 Lien, J., Gillian, N., Karagozler, M.E. et al. (2016). Soli: ubiquitous gesture sensing with millimeter wave radar. *ACM Transactions on Graphics* 35 (4): 1–19.

19 Manjunath, A., Liu, Y., Henriques, B., and Engstle, A. (2018). Radar based object detection and tracking for autonomous driving. *2018 IEEE MTT-S International Conference on Microwaves for Intelligent Mobility (ICMIM)*.

20 Dong, X., Shen, J., Wang, W. et al. (2018). Hyperparameter optimization for tracking with continuous deep Q-learning. *2018 IEEE/CVF Conference on Computer Vision and Pattern Recognition*. IEEE.

21 Gu, S., Lillicrap, T.P., Sutskever, I., and Levine, S. (2016). Continuous deep Q-learning with model-based acceleration. *CoRR*.

22 Servadei, L., Zheng, J., Arjona-Medina, J. et al. (2020). Cost optimization at early stages of design using deep reinforcement learning. *Proceedings of the 2020 ACM/IEEE Workshop on Machine Learning for CAD*, inser. MLCAD '20. ACM.

23 Bai, J., Li, S., Huang, L., and Chen, H. (2021). Robust detection and tracking method for moving object based on radar and camera data fusion. *IEEE Sensors Journal* 21 (9): 10–761–10–774.

24 Gao, X., Xing, G., Roy, S., and Liu, H. (2019). Experiments with mmWave automotive radar test-bed. *2019 53rd Asilomar Conference on Signals, Systems, and Computers*, pp. 1–6.

25 Will, C., Vaishnav, P., Chakraborty, A., and Santra, A. (2019). Human target detection, tracking, and classification using 24-GHz FMCW radar. *IEEE Sensors Journal* 19 (17): 7283–7299.

26 Santra, A. and Hazra, S. (2020). *Deep Learning Applications of Short-Range Radars*. Artech House.

27 Rohling, H. (1983). Radar CFAR thresholding in clutter and multiple target situations. *IEEE Transactions on Aerospace and Electronic Systems* AES-19 (4): 608–621.

28 Ester, M., Kriegel, H.-P., Sander, J., and Xu, X. (1996). A density-based algorithm for discovering clusters in large spatial databases with noise. *Proceedings of the Second International Conference on Knowledge Discovery and Data Mining*, ser. KDD'96, AAAI Press, pp. 226–231.

29 Wan, E. and Van Der Merwe, R. (2000). The unscented Kalman filter for nonlinear estimation. *Proceedings of the IEEE 2000 Adaptive Systems for Signal Processing, Communications, and Control Symposium (Cat. No.00EX373)*, pp. 153–158.

30 Kalman, R.E. (1960). A new approach to linear filtering and prediction problems. *Journal of Basic Engineering* 82 (1): 35–45.

31 Wan, E. and Merwe, R.V.D. (2000). The unscented Kalman filter for nonlinear estimation. *Proceedings of the IEEE 2000 Adaptive Systems for Signal Processing, Communications, and Control Symposium (Cat. No. 00EX373)*. IEEE, pp. 153–158.

32 Vaishnav, P. and Santra, A. (2020). Continuous human activity classification with unscented Kalman filter tracking using FMCW radar. *IEEE Sensors Letters* 4 (5): 1–4.

33 Roth, M., Hendeby, G., and Gustafsson, F. (2014). EKF/UKF maneuvering target tracking using coordinated turn models with polar/cartesian velocity. *17th International Conference on Information Fusion (FUSION)*, pp. 1–8.

34 Chen, Z., Heckman, C., Julier, S., and Ahmed, N. (2018). Weak in the NEES?: Auto-tuning Kalman filters with Bayesian optimization. *2018 21st International Conference on Information Fusion (FUSION)*. IEEE.

35 Zimmermann, K., Svoboda, T., and Matas, J. (2007). Adaptive parameter optimization for real-time tracking. *2007 IEEE 11th International Conference on Computer Vision*. IEEE, pp. 1–8.

36 Madrigal, F., Maurice, C., and Lerasle, F. (2019). Hyper-parameter optimization tools comparison for multiple object tracking applications. *Machine Vision and Applications* 30 (2): 269–289.

37 Ng, A.Y. and Russell, S.J. (2000). Algorithms for inverse reinforcement learning. *Proceedings of the 17th International Conference on Machine Learning*.

38 Levine, S. (2018). Reinforcement learning and control as probabilistic inference: tutorial and review. *CoRR*.

39 Liu, H., Zhang, H., and Mertz, C. (2019). DeepDA: LSTM-based deep data association network for multi-targets tracking in clutter. *2019 22nd International Conference on Information Fusion (FUSION)*.

40 Vaidehi, V., Chitra, N., Krishnan, C., and Chokkalingam, M. (1999). Neural network aided Kalman filtering for multitarget tracking applications. *Proceedings of the 1999 IEEE Radar Conference. Radar into the Next Millennium*. IEEE.

41 Shichao, X., Qun, Z., Ying, L., and Kaiming, L. (2020). Reinforcement learning based waveform design for cognitive imaging radar. *2020 IEEE 3rd International Conference on Electronic Information and Communication Technology (ICEICT)*. IEEE.

42 Martone, A.F., Sherbondy, K.D., Kovarskiy, J.A. et al. (2020). Metacognition for radar coexistence. *2020 IEEE International Radar Conference (RADAR)*. IEEE.

43 Gaafar, M., Shaghaghi, M., Adve, R.S., and Ding, Z. (2019). Reinforcement learning for cognitive radar task scheduling. *2019 53rd Asilomar Conference on Signals, Systems, and Computers*. IEEE.

44 Meng, F., Tian, K., and Wu, C. (2021). Deep reinforcement learning-based radar network target assignment. *IEEE Sensors Journal* 21 (14): 16315–16327.

45 Ahmed, A.M., Ahmad, A.A., Fortunati, S. et al. (2021). A reinforcement learning based approach for multi-target detection in massive MIMO radar. *IEEE Transactions on Aerospace and Electronic Systems* 57 (5): 2622–2636.

46 Selvi, E., Buehrer, R.M., Martone, A., and Sherbondy, K. (2018). On the use of Markov decision processes in cognitive radar: an application to target tracking. *2018 IEEE Radar Conference (RadarConf18)*. IEEE.

47 Ziebart, B.D., Maas, A.L., Bagnell, J.A., and Dey, A.K. (2008). Maximum entropy inverse reinforcement learning. In: *Proceedings of the 23rd AAAI Conference on Artificial Intelligence, AAAI 2008, Chicago, IllinWangois*, USA, 13–17 July 2008 (ed. D. Fox and C.P. Gomes). AAAI Press.

48 Kingma, D.P. and Ba, J.A. (2014). A method for stochastic optimization. *3rd International Conference for Learning Representations*, San Diego, 2015. *arXiv:1412.6980*.

5

Cross-Modal Learning

After reading this chapter, you should be able to

- Have a good overview of multimodal deep learning.
- Apply different multimodal deep learning approaches for radar use-cases.
- Design and implement cross-modal deep learning techniques in radar solutions.

5.1 Introduction

Before delving into cross-modal learning, let us take a brief look into the multimodal deep learning (MMDL) space of which cross-modal learning is a subclass. A deep learning approach that involves using single-modal stream as input in the learning process is called unimodal, whereas one that involves multimodal stream as input is called MMDL. MMDL has gathered significant amount of traction recently in both academic and industrial research, owing to its resemblance to how we perceive the world around us by employing multisensory input, i.e., hear, touch, see, smell, and taste. MMDL finds its application across various domains and use-cases. As covered in the review paper [1] by Summaira et al. many proposed solutions use MMDL approach to generate image description where for a given input image the proposed model should be able to generate visual descriptions in the form of text [2–4]. MMDL is also used for video description task, where textual descriptions are formulated for a given input video and can find wide range of use-cases such as video surveillance, subtitle generation, and sign language generation [5–7]. In use-cases such as speech generation or text-to-speech (TTS), both modalities (audio waveform and corresponding text) are used in a MMDL approach [8–10]. Visual question answering (VQA) is probably one of the most intuitive idea in terms of multimodal input where the model is able to comprehend questions and generate answers by amalgamating question

Methods and Techniques in Deep Learning: Advancements in mmWave Radar Solutions, First Edition.
Avik Santra, Souvik Hazra, Lorenzo Servadei, Thomas Stadelmayer, Michael Stephan, and Anand Dubey.
© 2023 The Institute of Electrical and Electronics Engineers, Inc. Published 2023 by John Wiley & Sons, Inc.

text and input image to which the question is tied to. In the next paragraph, a brief introduction to one such VQA [11, 12] approach is described to give readers an idea of how such a task can be addressed by leveraging a MMDL approach. The readers are not expected to have a complete in-depth understanding of the underlying architecture of the model discussed in the next paragraph but rather just have an idea of how different modalities are tackled and used, which can be helpful in comprehending detailed examples discussed later in the chapter.

VQA machines [13] as proposed by Wang et al. uses image, question, and facts related to the image as input to predict answer along with reasons for making the prediction. They employ hierarchical question encoding to generate an embedding for the questions where the words are represented in an one-hot encoding scheme and further fed to 1D convolution layers with varying filter size(unigram, bigram, and trigram) followed by pooling layers to generate phrase-level features. Finally, a long short-term memory (LSTM) layer encodes the phrase-level features to question-level features. The input image is resized and split into 14×14 regions which are extracted from last pooling layer of a VGG -19 or Res-Net-100 which is further embedded using a learned embedding weight and represent the image features. Inspired by the way of encoding facts in a knowledge graph, the authors propose a triplet format (subject, relation, object) and learnable weights W_n for each one-hot encoded element in the triplet. Different visual architectures are used to generate encoded facts list and can be easily extended to use any visual model which can extract facts more important for a given task. Sequential coattention is used to learn weights for different features (Question, Facts, Image) and the weighted features are fed to an multi-layer perceptron (MLP) which performs multiclass prediction where each distinct answer is a class.

Based on different learning approaches, MDL techniques can be classified as the following:

Self-Supervised MMDL where one modal's discriminative features can help other modal's learning. A generic example for such an approach would be using a input data stream with two modalities where one can act as a support modal S which is fed to a parameterized or nonparameterized function $f(x)$, e.g., clustering algorithms that generates distinctive features s_i for different possible classes. These features can be used as target prediction for the deep learning (DL) model that takes the other modal (input modal D) as input. The learning can be performed by maximizing the similarity between the output of the DL model d_i and s_i using a similarity function $S(x)$. The loss function can be formulated as follows:

$$L_{SS} = -\frac{1}{N}\sum_{i=1}^{N} S(d_i, s_i) \tag{5.1}$$

where N is the batch size.

Joint embedding learning where multiple modal's embedding's can be projected to a shared space. For example, an input data stream with two modals (D,S) are fed to two different DL embedding models to generate embeddings d_i and s_i. The loss function can be formulated as follows:

$$L_{JE} = \frac{1}{N} \sum_{i=1}^{N} \text{Dist}(d_i, s_i) \tag{5.2}$$

where $\text{Dist}(a, b)$ is a distance metric related to the space (e.g., Euclidean distance), and N is the batch size.

Multimodal input learning where the solution involves use of two modalities as input. One generic example would be using two input modals (D,S) and feeding them to two different DL networks and perform fusion of features d_i and s_i which act as input to a final prediction layer. Such an architecture can be either trained in an end-to-end fashion or use pretrained models to generate features, fuse them, and feed them to the prediction block which can be trained separately.

Cross-modal learning where there is knowledge distillation from one modal input to another. For example, a superior modal D is used to train a teacher network TN which is supposed to have better performance for the specific task. The other modal S is used as input to a student network SN which is trained by using distillation loss computed from predicted features s_i of SN and d_i generated from the pretrained TN. In inference mode, only modal S and SN is required which already has some degree of knowledge transfer from the superior modality.

The illustration in Figure 5.1 provides readers with an intuition of the above-mentioned learning algorithms. The chapter covers one example of each type of learning along with a radar use-case that uses cross-modal learning.

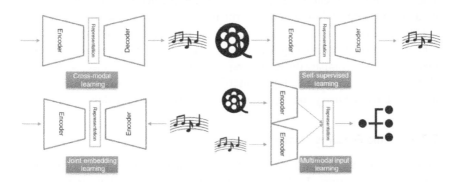

Figure 5.1 MMDL techniques.

5.2 Self-Supervised Multimodal Learning

In the chapter [14], the authors propose an interesting notion on how ambient sounds can be complementary to visual information and often relate to the same structure in space. For example, the sound of an engine can easily help us to distinguish between a fast-moving car or a heavily loaded truck based on its pitch. They present a model that can predict related audio cluster for a given input frame.

5.2.1 Generating Audio Statistics

The audio statistics are generated by splitting audio into 3.75s window and by following the audio texture model of McDermott which involves filtering the waveform with 32 band-pass filters lying within human cochlear frequencies and applying a Hilbert envelope, samples of which are increased by 0.3 power, and the envelope is resampled at 400 Hz. The averaged statistics of these subenvelopes over time is calculated by computing the mean and standard deviation of each frequency channel, Pearson's correlation between two channels, and the mean squared result of applying 10-band-pass filters on the channels with center frequencies ranging from 0.5 to 200 Hz which is equispaced on a logarithmic scale. In order to make the audio features(modulation power, energy, correlation, and marginal moments) invariant to gain, they are normalized and then rescaled inversely to the dimension of each feature. The sound texture vector for each image has a dimension of 502.

5.2.2 Predicting Sounds from Images

The author's motivation to predict sound from images lies in the idea of being able to learn visual features important for scene prediction. Only one frame is used instead of multiple as motion information that causes certain sound and may not always appear in the images. The task is formulated as a classification task by explicitly defining audio categories which act as labels for the images. This also makes it easier to compare the learned representations to that of object-scene models to prove that the learned representations are meaningful for the task. Vector quantization and binary coding scheme are used as labeling models.

5.2.3 Audio Features Clustering

The audio features are clustered by employing k-means, and the resulting clusters act as sound categories. Each audio feature is assigned a label according to the class of the closest centroid. Within a cluster, its seen that corresponding videos of the audios contain similar objects, while general scenes such as outdoor or wind blowing appear over all the clusters. The audio features that have a higher distance than the median distance to the centroid are removed.

5.2.4 Binary Coding Model

In binary coding model, principal component analysis (PCA) is employed on the audio features to generate top-30 components which are then thresholded to generate binary code. In the training, the model contains a sigmoid layer to predict the binary code, and cross entropy is used as the loss function.

5.2.5 Training

A subset of the Flickr video dataset which contains public videos was used with random selection of 10 frames per video. In total, 1.8 million training images were used which was fed to a CaffeNet Architecture-based model with batch normalization, batch size of 256, and Stochastic Gradient Descent as the optimizer. Figure 5.2 (adapted from Owens et al. [14]) depicts the overall solution architecture where an input image is fed to a convolutional neural network (CNN) that predicts the audio cluster from a set of audio clusters formed based on different scenes.

5.2.6 Results

A qualitative study on how the convolutional units are selective of specific objects and an analysis of their distribution are covered in the discussed literature

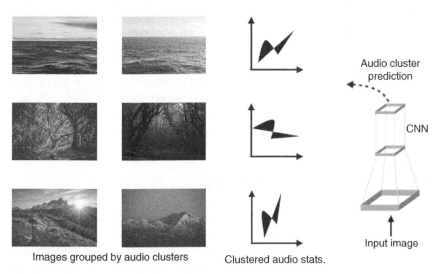

Images grouped by audio clusters   Clustered audio stats.

Figure 5.2 Proposed self-supervised learning-based architecture.

Figure 5.3 Example of neural visualization of the trained network for different groups (people, babies, waterfalls).

work. Furthermore, the quality of learned representations was comparable to representation learned through training methodologies which do not involve any human annotations. Figure 5.3 (adapted from Owens et al. [14]) represents example learned features by the network. One can notice how specific objects such as field, car, sea were extracted. The study contains additional material on qualitative and quantitative evaluation of the proposed solution.

5.2.6.1 Discussion

This form of training methodology can be really useful in cases where abundant data are available in the form of video but not annotated. Additionally, trying out different audio representations may yield different object detection. From a radar use-case perspective, one might use such a setup to perform crowd monitoring, for example where the radar data are preprocessed to form micro- and macro-range-Doppler image (RDI) which become inputs to a similar CNN network that predicts audio clusters representing densely crowded, sparsely crowded, or no crowd scenario.

5.3 Joint Embeddings Learning

In the chapter [15], the authors propose a novel technique to perform video-text retrieval task by exploiting multimodal cues in the video. This involves using multimodal features (visual representation, audio, and text) for a fusion methodology to learn joint embedding by employing multiple loss functions and a modified pairwise ranking loss. The proposed technique was evaluated on MSVD and

MSR-VTT datasets and was found to exhibit improved performance compared to other methodologies.

5.3.1 Feature Representations

Text: The text feature of joint embedding dimension of 1024 is generated using a gated recurrent units (GRUs) whose inputs are word embedding vectors of size 300 and is trained in an end-to-end fashion.

Object: In order to extract object information, a pretrained res-net-152 architecture trained on ImageNet dataset is employed which generates a encoded representation obtained from the penultimate fully connected layer.

Activity: In order to capture the temporal dynamics of the videos, a pretrained RGB -I3D is used which consists of 3D CNN architecture where the input (frames × image height × image width × channels) consists of 16 continuous frames and output being a 1024 dimension vector.

Audio: In order to encode corresponding audio for the given video, SoundNet [16] is used which produces a 1024 dimension vector.

5.3.2 Joint Embedding Learning

In order to preserve the semantic similarity between the video features and text features, the authors propose a modified pairwise ranking loss with use of semi-hard negative mining that allows effective embedding learning. The loss function enables a joint embedding learning characterized by Θ (Weights of the I3D model, SoundNet, and GRU). The proposed loss function can be formulated as follows:

$$\min_{\theta} \sum_{v} L(r_v)[a - S(v, t) + S(v, \hat{t}]_+ + \sum_{t} L(r_t)[a - S(t, v) + S(t, \hat{v}]_+ \tag{5.3}$$

where $L(.)$ is a weighting function for ranks and r_v and r_t are the corresponding ranking of the matching video and matching text, respectively. S is a cosine similarity function, and $\hat{t}$ and $\hat{v}$ are hardest negatives in each minibatch. Chapter 2 provides more comprehensive explanation about different forms of embedding learning.

5.3.3 Matching and Ranking

In a video-text retrieval task and vice versa, the most important information involves extracting the objects, actions, and scenarios which are achieved through the proposed pipeline depicted in Figure 5.4 (adapted from Mithun et al. [15]) As the name of the embedding spaces suggests, the object-text space maps object features with the corresponding text and the activity text space focuses

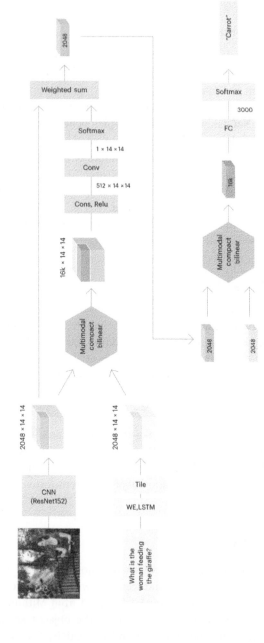

Figure 5.4 Overall architecture of the proposed solution for joint embedding learning.

more on mapping different events and actions to the description. Both of the embedding spaces are optimized using the proposed loss function (Eq. (5.3)). During inference, for a given text, the sum of similarity scores in both the spaces for all the videos in database is computed and ranked accordingly. The same can be done for an input video and ranking possible corresponding text.

5.3.4 Training Details and Result

The model is trained using Adam optimizer. The learning rate is set to 0.0002 for first 15 epochs which are further reduced by a factor of 10 for the next 15 epochs. The gradients are clipped if the L2 norm of the entire layer exceeds 2. In this chapter, the results of the proposed solution over other competitive solutions are displayed. R@1, R@5, R@10 indicates if the ground truth appears in top-1, top-5, and top-10, respectively, and the proposed solution demonstrates superiority overall as well as in most of the metrics. This chapter includes more detailed quantitative and qualitative results.

5.3.5 Discussion

In general, the proposal seems to very promising and provides an improved methodology for multimodal joint embedding learning. However, in certain cases, where the training data include very similar videos with few extreme outliers, the hard-negatives would lead to a very unstable model and might even collapse. One can use such a joint embedding for a radar solution, where the task is to have a very accurate keyword or voice recognition solution. The input for such a use case can be audio data and radar data which are fed to separate Sinc-net architecture (discussed in Chapter 3) and projected in a shared embedding space. During inference, just the radar data can be used to retrieve corresponding audio.

5.4 Multimodal Input

One of the biggest challenges in a multimodal input solution is defining a function that fuses multimodal representative vectors in a meaningful way. In this chapter [17], the authors propose a novel idea of generating an expressive function that takes in visual embedding vector along with a textual embedding vector to generate a joint representation vector. This is achieved by using multimodal compact bilinear (MCB) pooling. The proposed methodology is used for VQA task and achieves state-of-the-art results in the Visual 7W dataset and the VQA challenge.

5.4.1 Multimodal Compact Bilinear Pooling

In the task of VQA, a location in the input image or a possible answer is predicted where the inputs are an image I and a question Q. This can be formulated as follows:

$$\hat{a} = \text{argmax}_{a \in A} p(a|I, Q; W) \tag{5.4}$$

where A is the possible locations in an image or answers and W the parameters. MCB pooling allows to encode an image embedding and text embedding generated, for example by a CNN and an LSTM, respectively, in a meaningful representation that captures the underlying relationship between them and makes it easier for the classifier to learn. Bilinear models are able to linearize a matrix generated as a result of outer product of two vectors. However, in high-dimensional input vectors, this leads to very high number of learnable parameters. In order to counter the abovementioned challenge, count sketch algorithm is employed which projects the outer-product in a lower dimension without computing the product directly. Two vectors are initialized, $c \in \{-1, 1\}^n$ and $k \in \{1, ..., d\}^n$, where c contains either -1 or 1 for each index in the input, d is the dimension of the output, and y and k projects each index of input in index of output y. Both the vectors are initialized randomly from a uniform distribution, and y is initialized as a zero vector. Pham and Phag [18] demonstrated that the convolution of the count sketches expresses the outer product of two vectors, and as per convolution theorem, convolution in time domain can be expressed as inner product in frequency domain. This omits the requirement to compute the outer-product directly and is employed. The overall architecture of MCB is depicted in Figure 5.5 (adapted from Fukui et al. [17]).

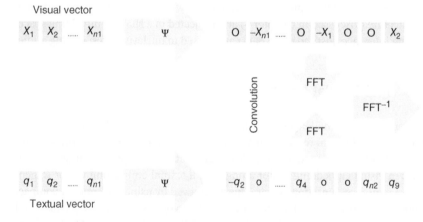

Figure 5.5 Multimodal compact bilinear block.

5.4.2 VQA Architecture

The goal is to be able to predict a possible answer from a large set of answers treated as classes, from a given input image and question. The image-embedding vector is generated by a pretrained Res-Net-152 trained on the ImageNet. The output of the penultimate pooling layer is used as the embedding vector having a dimension of 2048 on which L2 normalization is performed. The corresponding 2048 embedding vector for the input question is generated by concatenating the output of two LSTM layer each with 1024 units trained with one-hot encoded word tokens. Soft-attention is used to capture spatial information in the MCB block. The output of the MCB pooling is fed to two convolutional layers with a Relu activation and softmax activation, respectively. The output of the softmax activation produces a normalized attention map which is used to compute a weighted sum of the spatial vector (output of the image CNN) to generate the attended representation. This allows the architecture to focus on specific location in both representations. The generated embedding vectors are passed through an MCB block followed by L2 normalization and fully connected layer that maps it to a 3000 dimension classification layer. The proposed architecture is depicted in Figure 5.6 (adapted from Fukui et al. [17]).

This chapter contains additional similar architecture for VQA with multiple choices and visual grounding.

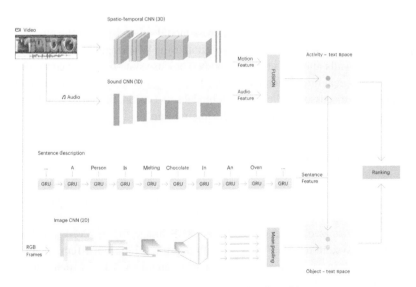

Figure 5.6 Overall architecture of proposed solution for VQA.

5.4.3 Training Details and Result

This chapter uses VQA dataset which contains approximately 200K MSCOCO images with three questions for each image and 10 answers for each question. A $2:1:2$ split is done for creating train, validation, and test dataset. Adam optimizer is employed with $\epsilon = 0.0007$, $\beta1 = 0.9$, $\beta2 = 0.999$ along with early stopping. Dropout is used in the LSTM and FC layers. In this chapter, the results of the proposed solution over other architectures demonstrate clearly its superior performance. This chapter includes more detailed quantitative and qualitative results on VQA and visual ground task.

5.4.4 Discussion

The proposed solution tactfully addresses the issue of multimodal outer-product without requirement of a huge computation using MCB block. Such solutions are particularly interesting for developing embedded solutions where often there is a computation constraint. For a radar use-case, such an algorithm can be used for a radar–camera-based solution, where the inputs can be the camera image and radar RDI Images.

5.5 Cross-Modal Learning

In this chapter [19], the authors leverage cross-modal supervision for the purpose of building an accurate pose estimation solution using Wi-Fi signals that can penetrate through walls and reflect human body. They use state-of-the-art computer vision model to generate skeletons and use them as ground truth for developing the solution. As discussed earlier, in cross-modal supervision, the teacher network (camera-based model) is only required during training.

5.5.1 Data Acquisition

The system relies on transmitting frequency modulated continuous wave (FMCW)-based low-power RF signals (complex) using a vertical and horizontal antenna. The setup allows for range and angular separation between the reflecting object. A synchronized camera provides an RGB image for the corresponding time-step. The system has an FPS of 30, with depth resolution of 10 cm, angular resolution of 15 in each direction, and a wavelength of 5 cm which causes reflection in a case of human body.

5.5.2 Cross-Modal Learning for Keypoint Detection

Cross-modal learning is employed to transfer the knowledge of a vision teacher model T, which generates the salient skeleton keypoints of human poses, each relating to a limb to a student network S which takes in RF signals as input and tries to reconstruct back the skeleton keypoints. This chapter uses partial affinity field (PAF) network [20] to generate 14 keypoints and confidence maps relating to neck, head, elbows, shoulders, knees, hip, ankles, and wrists. The vertical and horizontal RF signals are passed through individual encoders, outputs of which are concatenated channelwise and fed to a decoder that reconstruct the confidence maps. The loss function is defined as the summation of binary cross entropy of each pixel in the confidence maps and can be formulated as follows:

$$L = -\sum_{k}\sum_{x,y} S_{x,y}^{k} \log T_{x,y}^{k} + (1 - S_{x,y}^{k}) \log(1 - T_{x,y}^{k}) \tag{5.5}$$

where k is the confidence map and (x,y) the pixel location. In order to counter the problem of missing limbs in a single frame of RF signal caused due to its low-spatial resolution, a sequence of 100 (3.3s) frames is used. The decoder is designed in such a fashion that its outputs equal number of confidence maps. This enables the model to capture and aggregate information over multiple frames while being able to output confidence maps for each input frame. The RF encoding layer is built up by using 10 spatiotemporal convolution blocks to be able to learn salient features and be invariant in both space and time. The kernel size is of $9 \times 5 \times 5$ with stride of $1 \times 2 \times 2$ in every alternate layer. Relu is used as the activation function for all the layers. The decoder is made of four spatiotemporal convolution layers with kernel size of $3 \times 6 \times 6$ and fractionally stride of $1 \times 0.5 \times 0.5$ for all the layers except in the last one which has fractionally stride of $1 \times 0.25 \times 0.25$. Parametric Relu activation function is used for all the layers except for the output layer where sigmoid activation function is used. The complex values of RF signals are treated as two real channels by splitting the real and imaginary values. The overall architecture is depicted in Figure 5.7 (adapted from Zhao et al. [19]).

5.5.3 Training Details and Results

The dataset used for training and testing is composed of synchronized data of RF signal and corresponding camera image collected in 50 different scenarios and for a total duration of 50 hours. Different activities such as jogging, sitting, reading, and other natural activities were performed. On average, each frame contains 1.64 person and maximum of 14 persons. In order to be able to test partially occluded (by furniture, wall, or other static objects) scenarios, an eight camera-based 3D

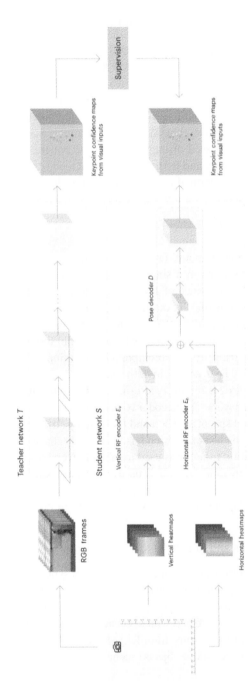

Figure 5.7 Overall architecture of the proposed solution for cross-modal Learning.

pose estimation algorithm to generate the pose and project into the view of camera is attached with the radio. This additional test setup data were only used for testing. A 70%–30% train-test split was done on the fully visible dataset and the test set additionally contained all the through-the-wall scenarios. The proposed solution is compared with state-of-the-art vision pose model. OpenPose and average precision over the object keypoint similarity (OKS) is used as a metric for evaluation. Furthermore, $AP^{0.5}$, which is average precision when OKS is 0.5 and is regarded as loose pose match, and $AP^{0.75}$, when OKS is 0.75 and is regarded as strict pose match, are used as additional evaluation metrics. The comparative results between the outputs of RFPose which takes in only RF signals as input and OpenPose which takes in RGB images as input is reported in this chapter, where for visible scenarios, the solutions show similar results, but what is most impressive is in through-walls scenarios where RF-pose is capable of performing without significant drop in accuracy compared to that of visible scenarios. This chapter contains further analysis of model and failed scenarios.

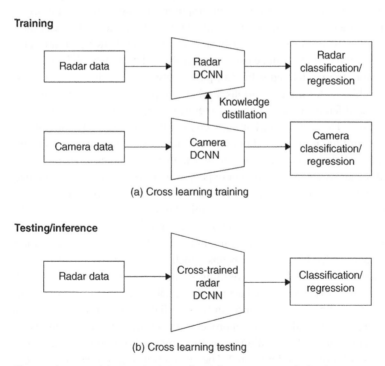

Training

(a) Cross learning training

Testing/inference

(b) Cross learning testing

Figure 5.8 (a) Training methodology involving knowledge distillation, (b) inference/testing mode.

5.5.4 Discussion

The proposed solution demonstrates a novel way of cross-modal learning in the form of a teacher–student network. Such a solution can be used to allow an inferior sensor in terms of resolution, for example (student) learn better features from a superior sensor (teacher) and during inference can deliver improved performance. The people counting solution using radar which is described in the next section leverages a similar learning setup (Figure 5.8).

5.6 Application: People Counting

People counting has been a core research topic across various domains in both industry and academia due to its wide range of use-cases. Accurate people counting can contribute to reducing energy consumption when integrated to systems such as HVACs, lights, and other consumer devices where the system's energy consumption can be controlled in proportion to the count of number of people in the area. Generating people count statistics for large, open areas such as malls, cinemas, airports can help businesses to make better planning and optimize workforce. It also can be used as a warning solution, where the number of people more than a defined value can pose safety issues such as in elevators, transport vehicles. Multiple solutions have been proposed in literature that involve using different sensors such as cameras, infrared, Lidar, radar, and other sensors. Camera-based solutions have proven to be the most accurate and reliable for people counting in indoor and outdoor settings and leveraging deep learning techniques, and such solutions have shown high performance even in highly populated scenarios where the number of people are in thousands. However, like any other sensor, a general camera-based solution comes with its own drawbacks, of which the most prominent ones are its limited to no performance in low-light scenarios and privacy concerns. On the other hand, radars are immune to such drawbacks but come with their own limitations of suffering from low-resolution data, multipath reflections and occlusions which result in missed detections in a conventional radar-based people counting solution. In our past work [21, 22], we proposed two novel radar-based deep learning solution for people counting in an indoor setup, where its able to accurately and reliably count number of people by leveraging cross-modal learning approaches to counter limitations discussed above. For simplicity and separation of the two proposed solutions, we represent the works as Solution 1 and Solution 2. The overall training and testing methodology of cross-modal learning through knowledge distillation is depicted in Figure 5.9.

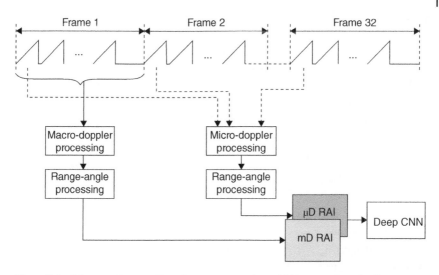

Figure 5.9 Micro- and macro-Doppler component-based RAI processing pipeline.

5.6.1 FMCW Radar System Design

In both the approaches, we use Infineon's BGT60TR13C radar system in the most common mode of having a stream of frequency chirps with short-ramp times and short delays between the chirps to make it power and computation efficient. The analog-to-digital converter (ADC) has 12-bit resolution that digitizes the data which are sent to PC via USB. The operating parameters of the radar for both the solutions are presented in Table 5.1.

5.6.2 Data Acquisition

For both the use-cases, we used our in-house data acquisition setup that includes synchronized radar-camera data collection which facilitates the cross-modal learning approach. The setup involves four synchronized cameras placed at the four corners of a room and at a height of 2 m. Each camera has a radar attached just below them.

5.6.3 Solution 1

5.6.3.1 Data Processing
Range-Angle Image Each raw ADC data frame is reshaped into a 2D matrix with a shape of number of chirps × number of samples. First, in order to get the range transformation, a 1D fast-Fourier transform (FFT) is taken across the samples for

Table 5.1 Operating parameters.

Parameters, symbol	Value
Ramp start frequency, f_{min}	60.5 GHz
Ramp stop frequency, f_{max}	61.5 GHz
Bandwidth, B	1 GHz
Range resolution, δr	15 cm
Number of samples per chirp, NTS	128
Maximum range, R_{max}	9.6 m
Sampling frequency, fs	2 MHz
Chirp time, T_c	64 μs
Chirp repetition time, T_{PRT}	400 μs
Maximum Doppler, v_{max}	3.125 m/s
Number of chirps, PN	64
Doppler resolution, δv	0.0977 m/s
Number of Tx antennas, N_{Tx}	1
Number of Rx antennas, N_{Rx}	3
Elevation θ_{elev} per radar	90°
Azimuth θ_{azim} per radar	130°

all the chirps. Another 1D FFT is taken for all the range bins to obtain Doppler information after performing mean subtraction across the chirps. The RDI maps formed as a result of these processing is further fed to a moving target indicator (MTI) filter that suppresses reflections from static objects in the room. In a similar fashion, we compute a micro-Doppler representation of the scene by forming a virtual frame made by concatenating first chirp from each frame for 32 consecutive frames and passing it through the same processing discussed above. The first RDI map allows us to capture major body motions, whereas the micro-RDI maps capture static person having micromotions due to breathing or small movements. Both RDI maps are further processed to form range angle images (RAIs) by using output of two antennas and feeding them to a digital beam-forming algorithm that uses weighted angle model that is formulated as follows:

$$z_{RAI}(r,\theta) = \sum_{v=-v_{max}}^{v_{max}} \sum_{j=1}^{N_{Rx}} z_{RDI}^{j}(r,v) e^{-j\frac{2\pi d^j \sin(\theta)}{\lambda}}$$

$$\forall -\frac{\theta_{azim}}{2} < \theta < \frac{\theta_{azim}}{2} \tag{5.6}$$

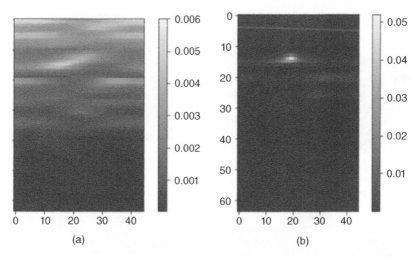

Figure 5.10 (a) Macro-Doppler RAI and (b) micro-Doppler RAI.

where θ_{azim} represents the half-power bandwith, z^j_{RDI} is the complex RDO from the jth receive channel, N_{Rx} is the number of receiver channels, and the outer summation is taken across the Doppler bins. Figures 5.9 and 5.10 represent the proposed radar data processing pipeline and sample macro–micro Doppler RAI generated for a static person, respectively.

Camera Data Processing The camera data are processed by using CSR-Net as proposed in paper [23]. The CSR-Net is able to perform accurate and robust people counting even in highly dense scenarios and can form a heat density map for a given input image. The proposed architecture reuses VGG-16 architecture's first 10 layers with three pooling layers. In order to have the same output dimension as of the input, bilinear interpolation is performed with a factor of 8 and is fed to an additional architecture consisting of 6 dilation convolution layers with dilation of 2 and number of feature maps being 512 for the first three and 256, 128, 64 for the last layers, respectively. The heat density maps are generated by employing a Gaussian kernel to blur each annotated heat in the input image and can be formulated as follows for a target object x:

$$F(x) = \sum_{i=1}^{N} \delta(x - x_i) \times G_{\sigma_i}(x); \quad \sigma_i = \beta d_i \tag{5.7}$$

where the average Euclidean distance of three nearest neighbor multiplied with a factor $\beta = 0.3$ represented by βd_i and is the standard deviation of the Gaussian kernel and $\delta(.)$ is the ground truth. The count of number of people can be performed by counting pixels above a defined threshold in a heat density map. In order to adapt and make CSR-Net robust for our indoor room scenarios, the last two layers

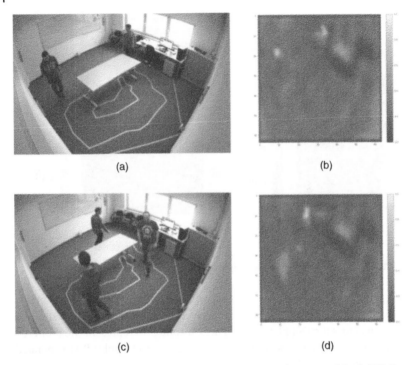

Figure 5.11 (a, c) Two and three person scenario camera images and (b, d) CSR-Net outputs, respectively.

of a pretrained CSR-Net is trained in supervised learning fashion with some annotated examples from our recordings. Sample results from the CSR-Net is displayed in Figure 5.11.

5.6.3.2 Learning Methodology

Inspired by the recent success of teacher–student networks for cross-modal learning, we build a DCNN -based autoencoder that takes in RAI images, both micro- and macro-RAIs fed as different channels, to reconstruct back heat density maps which are the output of the CSR-Net for the corresponding camera image. The learnt embedding should be representative enough to capture semantics derived from both radar and camera image which is superior in this case in terms of resolution and accuracy. In order to ensure a successful knowledge distillation from the camera network (teacher) to the radar DCNN (student), we propose a novel loss function that uses a combination of focal mean squared error (MSE) and cross-entropy. On successful training of the auto-encoder, the embedding layer output is passed to a fully connected layer with softmax activation and

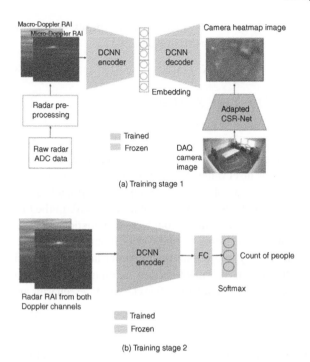

Macro-Doppler RAI
Micro-Doppler RAI
Camera heatmap image

DCNN encoder
DCNN decoder

Embedding

Radar pre-processing

Adapted CSR-Net

Trained
Frozen

Raw radar ADC data

DAQ camera image

(a) Training stage 1

Radar RAI from both Doppler channels

DCNN encoder → FC → Count of people

Softmax

Trained
Frozen

(b) Training stage 2

Figure 5.12 (a) Autoencoder learning. (b) Classification layer learning.

trained with ground truth people counting using binary cross-entropy loss. This allows us to have a cascaded networks which once trained can predict the number of people in a scenario by just taking RAI images as input. The two learning stages is depicted in Figure 5.12 The proposed radar DCNN network is detailed beneath.

Radar DCNN The radar DCNN follows an auto-encoder architecture and consists of three convolutional layers with Relu activation followed by pooling layer of factor 2 after each convolution layer in the encoder side. The decoder comprises of a similar architecture where pooling layers are replaced with upsampling layers with the same factor. In addition, the decoder part houses a convolution layer in the end with kernel size of 1×1 with sigmoid activation and 1 feature map. Remaining all convolution layers has a kernel size of 3×3 and 32 feature maps. Residual connection within the network is established by using add layers between the first two pooling and upsampling layers.

Classification First, the radar DCNN network is trained with RAI images as input and heat density maps from camera modality as reconstruction target. One key challenge for such a reconstruction is that most values in the density heat maps are near zero and very few really close to 1 which represents heads. We use the following proposed loss function to address this imbalance:

$$l_{MSE} = \frac{1}{N_0} w^0 \|X^0 - X_r^0\|^2 + \frac{1}{N_1} w^1 \|X^1 - X_r^1\|^2 \tag{5.8}$$

$$l_{CCE} = -\sum_{k \notin \Omega} (X^k)^{\gamma_0} \log(X_r^k) + \sum_{k \in \Omega} (1 - X^k)^{\gamma_1} \log(1 - X_r^k) \tag{5.9}$$

$$l_{total} = l_{CCE} + \lambda * l_{MSE} \tag{5.10}$$

where for input X and output X_r, the l_{MSE} involves using weighted summation parameterized by w^0 and w^1 for pixel values less and more than adaptive mean. The γ_0 and γ_1 controls the emphasizes of the same in the cross-entropy loss. The Ω includes pixels with value more than 0.4. The contribution of MSE in the total loss is limited by hyperparameter λ computed through cross-validation. Since we would want to emphasize more on learning the more critical but less-in-number pixel value, we choose values w^1, γ_1 more than their counter factor. The classification is done by using the output of trained encoder as input to a fully connected layer as described before. The classifier has five classes (1–4 scenarios and 4+ scenario).

5.6.3.3 Results
The dataset comprises of 59 720 frames with roughly 5000 frames per class. In the setup, 10 volunteers performed regular activities within the room and entered and left randomly. At any given time, there was not more than seven people in the room. The labeling involved using prediction from the CSR-Net followed by a manual verification. A 80%–20% train-test split was performed, and the test dataset only contained radar data. Two samples from the training dataset from displaying the reconstructed heatmap is shown in Figure 5.13. For the purpose of comparison with a unimodal approach, an exact same architecture was used where the auto-encoder had the same input and reconstruction target. In Figure 5.14, we display the confusion matrices for both the models. Clearly, the multimodal approach

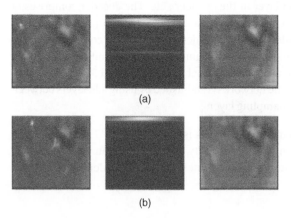

Figure 5.13 Order: ground-truth heatmap, macro-Doppler RAI, reconstructed heatmap. (a) Two-person scenario. (b) Three-person scenario.

(a)

(b)

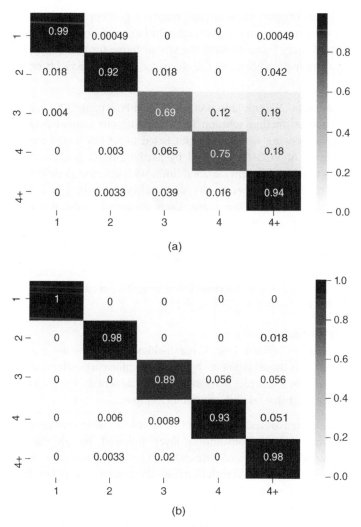

Figure 5.14 (a) Unimodal learning confusion matrix. (b) Cross-modal learning confusion matrix.

outperforms the unimodal approach and has an accuracy of 0.955 over 0.86. The encoder model along with the classification layer which is needed for inference has a size of 44 kb which makes it very feasible for an embedded deployment.

5.6.4 Solution 2

5.6.4.1 Data Processing
Range-Doppler Representation The raw ADC data is reshaped as a 2D matrix in a similar way as in solution 1, i.e., number of chirps × number of samples, and exact

same steps are followed to generate macro and micro range-Doppler representation. However, as in learning from our previous solution, we found that instead of using first chirp of every frame to form the virtual frame for micro-Doppler processing, using a summation across all the chirps for every frame yields better results.

Camera Data Representation We employ a pretrained OpenPose model [20], which is a multiperson pose estimation solution for processing our camera data. It comprises three-stage network where the input is fed to a CNN model that is made of first 10 layers of VGG-16, followed by partial affinity field network (PAF) which allows to relate different detected limbs with different persons and a confidence map refinement network. Greedy-matching algorithm is a used map of the PAFs and the confidence maps. Each generated confidence map represents the pixels where a particular limb might be present. We choose 13 such confidence maps representing 13 critical limbs and taken a mean of it to use it as an input to a triplet network with a 32 D linear embedding output. The triplet network is trained with all the people count classes and is used upon training to generate embedding for a given mean confidence map. The overall accuracy of the network with 70–30% train-test split was 98.75%.

5.6.4.2 Learning Methodology

In this solution, we try to perform knowledge distillation between the camera model (teacher) and radar model (student) by trying to enforce a similar embedding space that is learnt by the triplet network using the camera data to the radar encoder's embedding layer that takes in radar data as input.

Radar Encoder The radar encoder comprises of five Res-blocks which has residual connections involving a input-convolution layer followed by an output-convolution layer with kernel size of 3 times 3, a 1 times 1 convolution layer which helps in matching the input convolution layer dimension to that of output-convolution layer dimension for addition. The Res-blocks are followed by three fully connected layers with 128, 32 (same dimension as of the embedding layer in triplet network), and 7 (number of classes) hidden units. Relu activation is used for all the layers except the last two fully connected layers for which softmax activation is used.

Classification The input to the radar encoder have 12 channels representing real and imaginary value of micro- and macro-Doppler images for all three channels. The encoder is trained using softmax output of embeddings C^{emb} generated by the camera model for corresponding mean confidence maps which are used to compute Kullback–Leibler (KL)-divergence L_{KL} between it and the embedding layer E^{emb} of the radar encoder which is the second last layer. Also, the

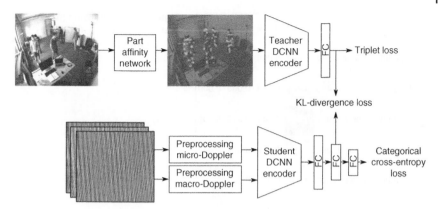

Figure 5.15 Proposed architecture.

ground-truth class count is one-hot encoded to compute the categorical cross-entropy L_{CCE} with the output of penultimate layer. Both of the losses are jointly optimized and can be formulated as follows:

$$L_{KL}(E^{emb} \parallel C^{emb}) = \sum_{x \in \mathcal{X}} E^{emb} \log \left(\frac{E^{emb}}{C^{emb}} \right) \tag{5.11}$$

where C is the number of classes $= 7$. During inference, index of the highest value in the last layer is predicted as people count. The overall proposed architecture is depicted in Figure 5.15.

5.6.4.3 Results

The dataset used for the evaluation of the proposed solution consists of 65 000 frames in the training set where people moved or stood still in a room and test set of 2600 frames recorded by a different radar sensor with a different placement to test the solution's generalization capability to different FOV. The classes are 0–6 people, and the results for a encoder model with knowledge distillation and without are depicted by the confusion matrices in Figure 5.16. We see a test accuracy of 71% and 58% with knowledge distillation and without, respectively. The most interesting learning is that while both models underperform in 2+ persons scenarios, the proposed solution exhibits a much more stable and close result. This difference is more clear when we look into the 2D uniform manifold approximation and projection (UMAP) of the embedding outputs from both the models and clearly see a better and more meaningful separation for the first.

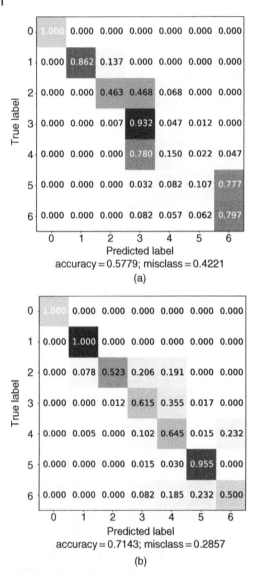

5.7 Conclusion

In this chapter, we went through the overall idea of multimodal learning and the value it brings in unimodal learning approaches. The multiple examples of multimodal learning that is covered can be stretched as a guidance and starting point for using different modalities for radar use-cases. The cross–learning-based radar solutions introduced for people counting use-case show the immense value such an approach brings for developing a stable and generalized solution while having the same computation and power cost in inference mode as that of a traditional approach.

5.8 Questions to the Reader

- What are the different types of modal present in a video recording?
- What are the key differences between multimodal and cross-modal learning approach?
- Define a radar-based use case for each type of MMDL learning approach discussed in the chapter?
- In a cross-modal learning approach, explain different approaches that can be followed for knowledge distillation?
- Describe a detailed framework for the following use-cases: The goal is to build a speaker verification solution where a radar device attached with a microphone facing the throat of the subject. Explain in terms of preprocessing, learning methodology, and bottlenecks if any.
- What are the key factors to be taken care of in the discussed people counting use-case to ensure a well-generalized model.
- In general, explain the different bottlenecks that may arise in MMDL approach.

References

1 Summaira, J., Li, X., Shoib, A.M. et al. (2021). Recent advances and trends in multimodal deep learning: a review. *CoRR*, vol. abs/2105.11087. https://arxiv.org/abs/2105.11087.

2 Wu, J. and Hu, H. (2017). Cascade recurrent neural network for image caption generation. *Electronics Letters* 53 (25): 1642–1643.

3 Guo, L., Liu, J., Yao, P. et al. (2019). MSCap: Multi-style image captioning with unpaired stylized text. *2019 IEEE/CVF Conference on Computer Vision and Pattern Recognition (CVPR)*. IEEE, June 2019. https://doi.org/10.1109/cvpr.2019.00433.

4 Feng, Y., Ma, L., Liu, W., and Luo, J. (2018). Unsupervised image captioning. https://arxiv.org/abs/1811.10787.

5 Liu, S., Ren, Z., and Yuan, J. (2021). SibNet: Sibling convolutional encoder for video captioning. *IEEE Transactions on Pattern Analysis and Machine Intelligence* 43 (9): 3259–3272. https://doi.org/10.1109/tpami.2019.2940007.

6 Feng, Y., Ma, L., Liu, W., and Luo, J. (2018). Unsupervised image captioning. *CoRR*, vol. abs/1811.10787. http://arxiv.org/abs/1811.10787.

7 Krishna, R., Hata, K., Ren, F. et al. (2017). Dense-captioning events in videos. *CoRR*, vol. abs/1705.00754. http://arxiv.org/abs/1705.00754.

8 Wang, Y., Skerry-Ryan, R.J., Stanton, D. et al. (2017). Tacotron: Towards end-to-end speech synthesis. *INTERSPEECH*.

9 Ö. Arık, S., Chrzanowski, M., Coates, A. et al. (2017). Deep voice: real-time neural text-to-speech. In: *Proceedings of the 34th International Conference on Machine Learning*, ser. Proceedings of Machine Learning Research, vol. 70 (ed. D. Precup and Y.W. Teh), 195–204 (06–11 August 2017). PMLR. https://proceedings.mlr.press/v70/arik17a.html.

10 Arık, S.O., Diamos, G., Gibiansky, A. et al. (2017). Deep voice 2: multi-speaker neural text-to-speech. *Proceedings of the 31st International Conference on Neural Information Processing Systems*, ser. NIPS'17. Red Hook, NY, USA: Curran Associates Inc., pp. 2966–2974.

11 Lobry, S., Marcos, D., Murray, J.J., and Tuia, D. (2020). RSVQA: Visual question answering for remote sensing data. *IEEE Transactions on Geoscience and Remote Sensing* 58: 8555–8566.

12 Yu, J., Zhu, Z., Wang, Y. et al. (2020). Cross-modal knowledge reasoning for knowledge-based visual question answering. *Pattern Recognition* 108: 107563.

13 Wang, P., Wu, Q., Shen, C., and van den Hengel, A. (2016). The VQA-machine: learning how to use existing vision algorithms to answer new questions. *CoRR*, vol. abs/1612.05386. http://arxiv.org/abs/1612.05386.

14 Owens, A., Wu, J., McDermott, J.H. et al. (2016). Ambient sound provides supervision for visual learning. *CoRR*, vol. abs/1608.07017. http://arxiv.org/abs/1608.07017.

15 Mithun, N.C., Li, J., Metze, F., and Roy-Chowdhury, A.K. (2018). Learning joint embedding with multimodal cues for cross-modal video-text retrieval. *Proceedings of the 2018 ACM on International Conference on Multimedia Retrieval*, ser. ICMR '18. New York, NY, USA: Association for Computing Machinery, pp. 19–27. https://doi.org/10.1145/3206025.3206064.

16 Aytar, Y., Vondrick, C., and Torralba, A. (2016). SoundNet: Learning sound representations from unlabeled video. https://arxiv.org/abs/1610.09001.

17 Fukui, A., Park, D.H., Yang, D. et al. (2016). Multimodal compact bilinear pooling for visual question answering and visual grounding. *Proceedings of the 2016 Conference on Empirical Methods in Natural Language Processing*. Austin, Texas: Association for Computational Linguistics, pp. 457–468. https://aclanthology.org/D16-1044.

18 Pham, N. and Pagh, R. (2013). Fast and scalable polynomial kernels via explicit feature maps. *Proceedings of the 19th ACM SIGKDD International Conference on Knowledge Discovery and Data Mining*, ser. KDD '13. New York, NY, USA: Association for Computing Machinery, pp. 239–247. https://doi.org/10.1145/2487575.2487591.

19 Zhao, M., Li, T., Alsheikh, M.A. et al. (2018). Through-wall human pose estimation using radio signals. *2018 IEEE/CVF Conference on Computer Vision and Pattern Recognition*, pp. 7356–7365.

20 Cao, Z., Hidalgo, G., Simon, T. et al. (2018). OpenPose: Realtime multi-person 2D pose estimation using part affinity fields. *CoRR*, vol. abs/1812.08008. http://arxiv.org/abs/1812.08008.

21 Stephan, M., Hazra, S., Santra, A. et al. (2021). People counting solution using an FMCW radar with knowledge distillation from camera data. *2021 IEEE Sensors*, pp. 1–4.

22 Aydogdu, C.Y., Hazra, S., Santra, A., and Weigel, R. (2020). Multi-modal cross learning for improved people counting using short-range FMCW radar. *2020 IEEE International Radar Conference (RADAR)*, pp. 250–255.

23 Li, Y., Zhang, X., and Chen, D. (2018). CSRNet: Dilated convolutional neural networks for understanding the highly congested scenes. *CoRR*, vol. abs/1802.10062. http://arxiv.org/abs/1802.10062.

6

Signal Processing with Deep Learning

After reading this chapter, you should

- Have an overview of different model-based approaches to incorporate expert knowledge in deep learning methodologies.
- Understand the advantages of including knowledge from signal processing methods in deep learning models and when to do so.
- Acknowledge the advantage of combining signal processing with deep learning for radar target segmentation.

6.1   Introduction

Machine learning and especially deep learning have proven to achieve state-of-the-art results on a wide range of different tasks. Breakthroughs in computer vision (CV), with convolutional architectures such as AlexNet [1] topping the ImageNet challenge, architectures such as Yolo enabling accurate real-time object-detection [2], and in natural language processing, with long short-term memories (LSTMs), and later transformers [3], enabling the use of context information over longer texts, have accelerated research in machine learning and deep learning. In many of these highly complex problems, machine-learning methods manage to exceed the performance of signal processing-based methods that require expert knowledge and typically rely on assumptions and approximations to reduce the complexity of the problem. Machine learning (ML)-based methods have also found their entrance in radar signal processing tasks, such as human target segmentation [4], people counting [5], gesture recognition [6], and hyperparameter estimation [7]. While showing promising results in most of these tasks, the data-driven nature of ML introduces some new issues.

Methods and Techniques in Deep Learning: Advancements in mmWave Radar Solutions, First Edition.
Avik Santra, Souvik Hazra, Lorenzo Servadei, Thomas Stadelmayer, Michael Stephan, and Anand Dubey.
© 2023 The Institute of Electrical and Electronics Engineers, Inc. Published 2023 by John Wiley & Sons, Inc.

Labeled data requirements that may not be given, overfitting on the training data distribution, failure or unpredictable behavior in unseen cases, data and computation requirements, and the issue of explainability due to the mostly black-box nature of modern neural networks. This chapter will focus on different approaches of combining expert or domain knowledge from signal processing with the flexibility and power of deep learning.

6.2 Algorithm Unrolling

Algorithm unrolling describes the technique of taking an iterative algorithm to solve a specific problem and then unrolling this algorithm for a specific number of iterations, whereby a neural network layer captures each step. This method was first proposed in [8] for sparse coding, demonstrating a better performance in fewer iteration steps. Since then, research interest in the possible applications of algorithm unrolling has led to publications in various fields of interest, such as super-resolution [9], image denoising [10], scene-flow estimation [11], and communications [12]. An overview of some different use cases of algorithm unrolling is given in [13]. Figure 6.1 shows the unrolling of a general algorithm as proposed in [8]. Here, x_0 is the initial state, representing raw data or some initial guess. The initial state x_0 is then fed to some algorithm h, which outputs an approximate solution y_0 for the first step. In an iterative algorithm, this approximate is then the input to h in the next step to achieve a better approximate solution. This is repeated for t times until we arrive at a satisfactory solution based on predefined criteria or for a specified number of steps t. We will better understand the concept of algorithm unrolling after going through the paper [8] as an example.

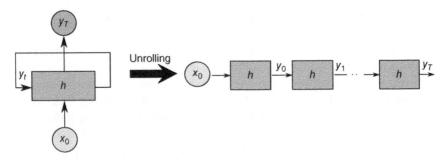

Figure 6.1 Visualization of an unrolled general algorithm as block diagram.

6.2.1 Learning Fast Approximations of Sparse Coding

The sparse coding problem is given by Eq. (6.3). The goal here is to find a latent representation $\mathbf{r}_i$ for each $\mathbf{x}_i$, such that $\mathbf{x}_i$ is mostly recoverable from r_i, i.e., the mean squared error between $\mathbf{x}_i$ and $\mathbf{Dr}_i$ is small, while being sparse.

$$\underset{\mathbf{D},r_i}{\operatorname{argmin}} \sum_{i=1}^{K} \|\mathbf{x}_i - \mathbf{Dr}_i\|_2^2 + \lambda \|\mathbf{r}_i\|_0 \tag{6.1}$$

The sparsity of r_i is ensured by the l_0-norm term in the optimization problem. The hyperparameter λ defines the weighting of the sparsity regularization term. $\mathbf{D} \in R^{n \times m}$ is an application-specific codebook, with the norm of its columns constrained to $\|d_i\|_2^2 \le 1$, to avoid the values in r_i becoming too small, that may either be defined or also part of the optimization problem. Solving Eq. (6.3) is NP-hard due to the l_0-norm [14]. Therefore, the l_0-norm is typically replaced with the l_1-norm as a convex relaxation:

$$\underset{\mathbf{D},r_i}{\operatorname{argmin}} \sum_{i=1}^{K} \|\mathbf{x}_i - \mathbf{Dr}_i\|_2^2 + \lambda \|\mathbf{r}_i\|_1 \tag{6.2}$$

The l_1-norm $\|\mathbf{r}_i\|_1$ of a vector $\mathbf{r}_i$ returns the sum of the absolute values of all its elements, and it was shown to exactly approximate the l_0-norm solution under certain conditions [15] while still achieving good results even if these conditions are not met [16]. If we assume that the codebook $\mathbf{D}$ is known, which is often given in the overcomplete case, where $m > n$, then we can write Eq. (6.2) for a single measurement vector $\mathbf{x}_i$ as follows:

$$\underset{\mathbf{r}}{\operatorname{argmin}} \|\mathbf{x} - \mathbf{Dr}\|_2^2 + \lambda \|\mathbf{r}\|_1 \tag{6.3}$$

A popular iterative way of solving the optimization problem of the above type is the iterative shrinkage-thresholding algorithm (ISTA). ISTA is a gradient-based method. The gradient of the quadratic part of Eq. (6.2) is $-2\mathbf{D}^T(\mathbf{x} - \mathbf{Dr})$. ISTA is then a simple algorithm with the following update step:

$$\mathbf{r}_{k+1} = S_\alpha(\mathbf{r}_k + 2t_k \mathbf{D}^T(\mathbf{x} - \mathbf{Dr}_k)) \tag{6.4}$$

where t_k is the step-size for the gradient update step, and $S_\alpha = (|\mathbf{r}_j| - \alpha)_+ \operatorname{sign}(\mathbf{r}_j)$ is a shrinkage function. $S_\alpha(\mathbf{r})$ shrinks any element $\mathbf{r}_j$ in $\mathbf{r}$ by a step-size of α toward 0 if $|\mathbf{r}_j| > \alpha$, and sets $|\mathbf{r}_j| = 0$, otherwise. With each step in the form of Eq. (6.4), the quadratic term is further minimized by following its negative gradient, while the regularizing l_1-norm term is kept small with the shrinkage operator S_α. In [8], they first reformulate the iterative update equation to

$$\mathbf{r}_{k+1} = S_\alpha(\mathbf{r}_k(\mathbf{I} - \mu_k \mathbf{D}^T\mathbf{D}) + \mu_k \mathbf{D}^T\mathbf{x}) \tag{6.5}$$

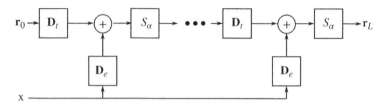

Figure 6.2 ISTA algorithm unrolled for L iterations.

where $\mu_k = 2t_k$. Then they substitute $\mathbf{D}_t = (\mathbf{I} - \mu_k \mathbf{D}^T \mathbf{D})$, and $\mathbf{D}_e = \mu_k \mathbf{D}^T$ to arrive at following expression:

$$\mathbf{r}_{k+1} = S_\alpha(\mathbf{r}_k \mathbf{D}_t + \mathbf{D}_e \mathbf{x}) \tag{6.6}$$

Figure 6.2 shows the diagram of the unrolled learned ISTA algorithm over L steps. In each step, the algorithm consists of two matrix-vector-multiplications, followed by a nonlinearity in S_α. Each such step can be directly realized by a neural network block consisting of two fully connected layers with linear activations and an addition of their outputs followed by a nonlinearity.

The full network consists of L concatenated ISTA-layers, as shown in Figure 6.3, where fully connected layers replace the mutual inhibition matrix $\mathbf{D}_t$ and $\mathbf{D}_e T$. The fully connected layers can now be trained in a supervised manner. For this, an optimal latent representation, or coded version for the input examples, is required. If given, the ISTA-layers in the network can be trained to approximate the desired code by minimizing the loss function shown in Eq. (6.7).

$$L(\mathbf{D}, \boldsymbol{\mu}) = \frac{1}{N} \sum_{i=1}^{N} \|\mathbf{r}_i^* - \mathbf{r}_{Li}(\mathbf{D}, \mu)\|_2^2 \tag{6.7}$$

Here, N is the size of one training-batch, $\mathbf{r}_i^*$ is the optimal code for the ith input example, and $\mathbf{r}_{Li}$ is the output code of the network for the ith example, produced after going through L ISTA-layers. The trainable parameters of the ISTA-layers can then be updated by minimizing the loss function in Eq. (6.7) via stochastic gradient descent and by backpropagating the gradients through the ISTA-layers. In practice, the codebook $\mathbf{D}$ may be fully trainable, or assumed to be partly known/parametric. Also, the weights representing the codebook $\mathbf{D}$ may be shared among the ISTA-layers, or can be different, $\mathbf{D}_k$, for each ISTA-layer, which generalizes the ISTA-algorithm.

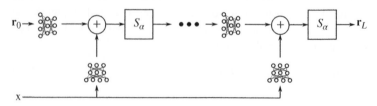

Figure 6.3 Learned ISTA algorithm L for multiple iterations.

6.2.2 Learned ISTA in Radar Processing

In radar signal processing, iterative gradient-based methods can be used to accurately estimate various parameters such as range, angle, or velocity of a target. Specifically, those methods, like ISTA, may be used for frequency estimation tasks when the received signal is somewhat sparse in its frequency representation. The overcomplete dictionary then consists of some M complex exponential monotones, as seen in Eq. (6.8).

$$\mathbf{D} = e^{j2\pi n/M}, \quad \text{with } n = 0, ..., M - 1 \tag{6.8}$$

In [17], they investigate the usage of learned ISTA for direction of arrival (DoA) estimation with radar antenna arrays while leveraging the special Toeplitz structure of the Gram matrix $\mathbf{D}^H\mathbf{D}$ in this specific case to reduce the computational complexity.

Estimating the DoA with radar antenna arrays is a frequency estimation problem. With a point-target reflecting the radar signal, under the far-field assumption, the angle information is encoded in the phase difference of the received target signal among the Rx antennas. Equation (6.9) shows the steering vector for a uniform linear array with N antenna elements.

$$\mathbf{a}(\phi) = e^{j2\pi f_c n \frac{\sin(\phi)d}{c}}, \quad n = 0, ..., N - 1 \tag{6.9}$$

Here, f_c is the carrier frequency, ϕ is the DoA of the signal, and d is the distance between the receiving antenna elements. The phase difference is given by $f_c n \frac{\sin(\phi)d}{c}$ and caused by the different propagation distance, Δd, among the antennas, as visualized in Figure 6.4.

The received radar signal can then be described as follows:

$$\mathbf{y}(t) = \mathbf{A}\mathbf{s}(t) + w(t) \tag{6.10}$$

where $\mathbf{s}(t) = [s_1(t), s_2(t), ..., s_N(t)]^T$ is the reflected signal originating from N targets at timestep t, $\mathbf{A} = [\mathbf{a}_1(\phi_1), \mathbf{a}_2(\phi_2), ..., \mathbf{a}_N(\phi_N)]$ is a matrix of steering vectors for each

Figure 6.4 Sketch of a uniform linear array (ULA) with three receivers, showing the different signal propagation paths.

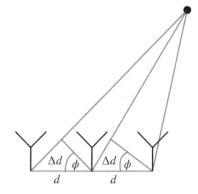

reflected signal, and $w(t)$ is some additive noise. The task of the DoA estimation may now be formulated as an optimization problem according to Eq. (6.2), where the codebook $\mathbf{D}$ consists of steering vectors, sampled over a discretized angle space, as seen in Eq. (6.9), and the objective is to find a combination of elements $\mathbf{a}_n(\phi_n)$ within the codebook, that best describes the received signal. Therefore, the DoA estimation problem may be tackled with sparse recovery methods, like ISTA, or learned ISTA, as described in Section 6.2.1. To reduce the computational complexity, the authors in [17] use the special Toeplitz structure of the Gram-matrix $\mathbf{D}^T\mathbf{D}$ and the mutual inhibition matrix $\mathbf{I} - \mathbf{D}^T\mathbf{D}$, built from the steering vectors. A complex Toeplitz matrix with a real diagonal has only $2M - 1$ degrees of freedom, as is visible in Eq. (6.11), compared to a normal square matrix with $M \times M$ degrees of freedom; therefore, the number of parameters for the layer representing the mutual inhibition matrix can be drastically reduced.

$$\mathbf{T} = \begin{bmatrix} x_0 & x_1 & \cdots & x_{(M-1)} \\ x_1 & x_0 & \cdots & x_{(M-2)} \\ \vdots & \vdots & \ddots & \vdots \\ x_{M-1} & x_{M-2} & \cdots & x_0 \end{bmatrix} \tag{6.11}$$

Also, by making use of the assumed Toeplitz structure, they can replace the matrix multiplication implemented by a fully connected layer representing the mutual inhibition matrix by a convolutional layer according to the relation shown in Eq. (6.12) and claim that this increases the learning efficiency and also the performance.

$$\mathbf{Tv} = \mathbf{x} \circledast \mathbf{v}, \quad \text{where } \mathbf{x} = [x_0, x_1, \ldots, x_{(M-1)}] \tag{6.12}$$

Additionally, since their input radar data is complex-valued, and the DoA information is encoded in the phase of the input signal, they use network operations that preserve the link between the real and imaginary part of the input and, therefore, the phase information. They then evaluate their proposed method with simulated data by training the learned ISTA-Toeplitz network on data created from five simulated, independent target sources at different angles, different noise levels, and the corresponding labeled angles. They show in their results that the learned ISTA and their learned ISTA-Toeplitz both perform better in terms of normalized mean squared error in fewer iterations. Additionally, they show that the complex learned ISTA-Toeplitz outperforms the complex learned ISTA even though it has fewer parameters.

6.3 Physics-Inspired Deep Learning

Deep learning methods are typically data-based. Therefore, supervised methods typically require a large amount of training data and corresponding labels depending on the methods used. Especially for many radar tasks, acquiring the necessary datasets may be infeasible or at least cost- or time-prohibitive. This section looks at different ways of simulating radar data with physics-based simulations and replacing these simulations with general adversarial networks. In [18], the authors compare various radar simulation methods and their feasibility for automotive radar target classification with convolutional networks. Typical targets that need to be classified in automotive radar applications are vehicles, cyclists, pedestrians, and animals. The usual features used for classifying these types of moving targets are micro-Doppler images [19], which can capture movement details such as arm-movement, or leg-movement, on top of the macromovements of the target. Three different types of physics-based simulations are as follows:

1. Primitive-based simulations.
2. Ray tracing simulations.
3. Full electromagnetic wave simulations.

Primitive-based simulations split the simulated target into independent shapes, whereby the total target return is given by the superposition of the individual shape returns. The received signal is then analytical calculated, typically under the far-field approximation and neglecting other effects like multi-path reflections or occlusions. Compared to full electromagnetic wave simulations, which are accurate but prohibitively expensive in terms of computational resources, ray tracing methods are still reasonably accurate but of potentially manageable complexity. Ray tracing methods simulate the wave propagation via multiple rays, originating from the radar sensor into different directions, and rebouncing upon hitting the first target. A trade-off between the accuracy of the simulation and computation time is managed by setting the number of emitted rays and the number of simulated bounces for each ray. The authors in [18] then use an adjusted ray-tracing-based method to simulate pedestrians, cyclists, dogs, and cars. They show that the simulated micro-Doppler data looks visually similar to measured data and then use it to train a convolutional neural network (CNN) on the classification task. While the trained network obtains an accuracy close to 100% on the simulated validation data, testing with real measured data was not done. The authors of [20] argue that any model-based simulation will not be able to cover all real effects like different noise levels, nonlinearities in the radar hardware, coupling, and signal dispersion. They, therefore, propose a data-driven simulation method by using generative adversarial networks and conditional variational autoencoders to create new training examples. Figure 6.5 shows a sketch of a general adversarial network.

The basic idea of a generative adversarial network is that the generator improves in creating samples from the training data distribution by competing with the

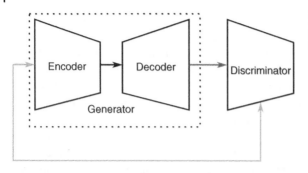

Figure 6.5 Sketch of a generative adversarial network.

discriminator. The discriminator gets real data samples (light gray in Figure 6.5) or generated data samples (dark gray in Figure 6.5) and tries to predict whether or not the sample was real or generated. In [20], they use a modified version of a generative adversarial network, namely an auxiliary classifier general adversarial network, which allows for conditioning on external labels for improved image quality. However, the authors in [20] find that some of the generated images contain unreasonable or physically impossible data points, which would degrade the training performance. To that end, they devise a way to filter out a number of these impossible data points and then use the other generated examples to help in the training of a human motion classification network. The results show that while real data are superior to those data points generated by the generator, if only a few data are available, then pretraining the network with synthetic data can boost the validation accuracy significantly.

6.4 Processing-Specific Network Architectures

Another way to inject domain knowledge into the neural network training process is to adjust the network architecture according to some known properties of the input features. Some well-known general architectures take advantage of some domain-specific input information, such as convolutional neural networks, recurrent neural networks, and graph neural networks. Deep convolutional neural networks are often used to extract local patterns of image-like inputs and preserve their spatial information. Recurrent neural networks were specifically designed to extract time information from the input data. Further, gated recurrent units and long short-term memory networks are evolutions of recurrent neural networks to better handle and keep valuable state information over longer time sequences. Graph neural networks work well with point-cloud-like input data, with nodes, or verticles, with their corresponding features, and edges, or links, that describe the (e.g., spatial) relation between some nodes. Other possibilities of adding domain knowledge to the network architecture include the usage of parametric layers [21] or separate network paths for different input features, whose information may be fused at some point in the network [21]. This section will go into some exemplary research where radar-specific domain knowledge was used to arrive at better-suited network architectures.

As the main example, this paper [22] takes advantage of ad hoc deep learning layers for processing the Doppler spectrograms. To this end, using deformable convolutions, the network can adapt to the highly deformed Doppler spectrograms. This has been particularly relevant in fields such as elderly fall detection in terms of application. Figure 6.6 presents an example of deformed spectrograms in a trip-and-fall backward scenario.

To process those spectrograms, in this contribution [22], not only deformable convolutions are applied but also a new loss formulation, the *effective loss*. The effective loss is expressed as shown in Eq. (6.13):

$$L(\mathbf{p}, \mathbf{y}) = -(1 - y)p^{\gamma} \log(1 - p) - y\alpha(1 - p)^{\gamma} \log(p)$$
$$+ y(1 - \alpha)\frac{1}{nk} \sum_{i=1}^{n} (x_i - m)^T (x_i - m) \tag{6.13}$$

Here, n is the number of samples in a minibatch, k is the number of frames, m is the mean of the fall-motion class, x_i is the single input frame, p is the probability of a class probability in a binary classification, and γ is a parameter which smoothly adjusts the rate at which easy examples are down-weighted. The effective loss combines the *focal loss* [23] and 1-class *contrastive loss* [24] for the fall motion class and helps in not only creating a hyperplane to classify the binary classes but also projecting the data into an embedding space wherein all fall motion data appear close together in a cluster. This encourages the model to generalize to unseen fall motions as they would be projected very close to the fall cluster in the embedding space, and the classification hyperplane would correctly classify the fall motion.

Figure 6.6 Doppler spectrogram of trip-and-fall forward of an elderly person.

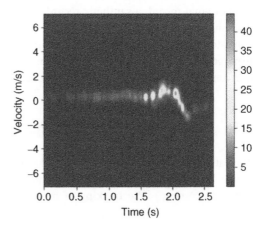

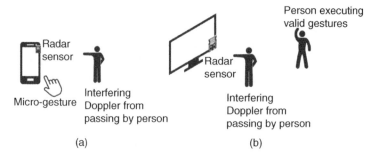

Figure 6.7 Applications of gesture recognition algorithms: (a) micro- and (b) macro-gesture application.

An additional contribution exploits the temporal dimension of the radar data in order to perform a gesture recognition operation [6]. The application of this method can be seen in Figure 6.7.

Here, the temporal sequence modeling is done by using an *LSTM* [25] layer that accepts a flattened output from the All-CNN [26] of sequence length of 100 frames. This model merges an intense hierarchical optical feature extractor (All-CNN) with a model which can absorb the knowledge of temporal dynamics for tasks involving sequential data. Additionally, a new data augmentation method has been proposed, which enhances the model's performance. Here indeed, the range-Doppler input images are augmented. Since the data are collected by a limited number of users, some variance is added in the range-Doppler images (RDIs) for training. The data augmentation technique proposed increases of the dataset size by creating synthetic images with variance to the original RDIs to address the generalization concerns. Initially, for each gesture class and for each time step, a mean range-Doppler image is formed from all channels. Next, for each original record, a synthetic record is formed by generating values for each pixel in time step t RDI by drawing values from a normal distribution with mean equal to the original Range-Doppler map at that time step and variance drawn from gesture class variation. This accounts for the time variations with which different individuals make the same gesture.

6.5 Deep Learning-aided Signal Processing

Typically, a lot of signal processing tasks may be split into different subtasks, which, put together, accomplish the overall goal. A good example of this in radar signal processing is the tracking of multiple static and moving targets within

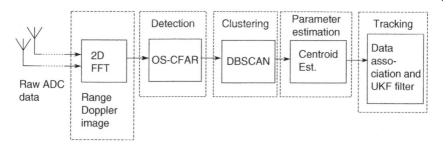

Figure 6.8 Traditional processing pipeline for radar target tracking.

the field of view of the radar. Figure 6.8 shows a potential signal processing chain, similar to that described in [27], for the task of multitarget-tracking with a radar sensor in the form of a block diagram. In the displayed processing chain, the raw digitized input data, containing information about potential targets, is first transformed to frequency domain, specifically, to an RDI. Then potential targets are detected within the RDI by a peak-finding and a clustering algorithm. Here, those blocks are realized with ordered statistics-constant false alarm rate (OS-CFAR) for peak-finding and density-based spatial clustering of applications with noise (DBSCAN) as clustering algorithm. After clustering, the cluster centers have to be estimated, which will then be used over multiple timesteps as inputs for some track association and track filtering block. For every single block in the whole tracking chain, a multitude of different signal processing algorithms, based on different assumptions and with various trade-offs, exist. Here, we look at the detection part of the chain, consisting of peak-finding and clustering, as an example. There has been extensive research for implementing both of these signal processing blocks. Some peak-finding methods commonly used in radar signal processing are simple thresholding based on the average receiver noise level, constant false alarm rate detectors (CFARs), which have adaptive thresholds based on the local noise levels, like cell averaging CFAR, and ordered statistics CFAR (OS-CFAR), or other methods that use additional domain knowledge, such as minimum peak prominence, minimum peak separation, or a predefined number of peaks. There are also various clustering methods, such as hierarchical-based methods, distribution-based methods, and density-based methods. A good overview of different clustering algorithms is given in [28].

To illustrate the method and the advantages of substituting these signal processing blocks, we take a closer look at the research in [4] in this section. Here, the authors propose a deep neural network-based method of segmenting multiple human targets in RDIs created from a frequency-modulated continuous-wave (FMCW) radar sensor. The chirp settings for the radar sensor are shown in Table 6.1.

Table 6.1 Operating parameters.

Parameters, symbol	Value
Ramp start frequency, f_{min}	58 GHz
Ramp stop frequency, f_{max}	62 GHz
Bandwidth, B	4 GHz
Range resolution, δr	3.75 cm
Number of samples per chirp, NTS	256
Maximum range, R_{max}	4.8 m
Sampling frequency, fs	2 MHz
Chirp time, T_c	261 μs
Chirp repetition time, T_{PRT}	520 μs
Maximum Doppler, v_{max}	4.8 m/s
Number of chirps, PN	32
Doppler resolution, δv	0.3 m/s
Number of Tx antennas, N_{Tx}	1
Number of used Rx antennas, N_{Rx}	3
Elevation θ_{elev} per radar	70°
Azimuth θ_{azim} per radar	70°

With these parameters, the size of one raw radar data frame is $3 \times 256 \times 32$, given by the number of antennas, the number of analog-to-digital conversion (ADC) samples per chirp, and the number of chirps. The input to the detection block is then of size 128×32 after two 1D fast-Fourier transforms (FFTs) along the sample and the chirp axis, with the respective mean subtractions before each FFT to remove static targets and the Tx-Rx-leakage, and maximum ratio combining over the antenna dimension to increase the signal-to-noise ratio. The objective of the work was then to replace the detection block, consisting of OS-CFAR and DBSCAN, with a Deep Residual U-Net, and show that the neural network has multiple advantages in terms of detecting partly occluded targets and rejecting ghost targets caused by multipath reflections.

Figure 6.9 shows the signal processing chain, going from RDIs after mean-subtraction to segmented RDIs, where the clusters represent the detected targets. The two blocks in the presented detection chain, OS-CFAR, and DBSCAN are widely used algorithms for these tasks. OS-CFAR is a sort of peak finding method that utilizes a sliding window and some noise statistics to adaptively threshold the RDIs. Figure 6.10 visualizes the 2D OS-CFAR method. There are three main components, the cell under test (CUT), the sliding window, with its size being

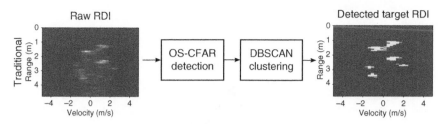

Figure 6.9 Traditional signal processing chain for target detection.

Figure 6.10 Visualization for th OS-CFAR process, with the cell under test, the guard window, the reference window, and the reference cell.

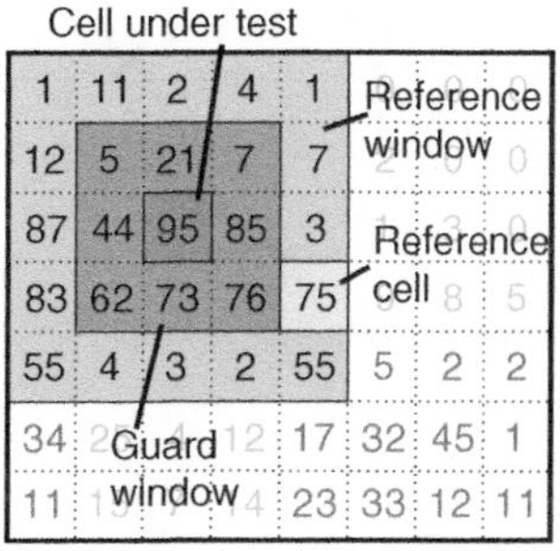

determined by the reference window size and the guard size, and the parameter k, which determines which cell in the reference window is used to estimate the noise level and calculate the threshold for the CUT.

After identifying potential targets via OS-CFAR, the detections need to be clustered, as human targets are spread across range and Doppler, and to reject outliers. The density-based algorithm DBSCAN is described in Figure 6.11. In DBSCAN, there are two free hyperparameters, namely the minimum number of neighbors N and the maximum neighbor distance d. As visualized in different colors, the algorithm clusters the available points into three different groups, core points (circles), edge points (crosses), and noise points (squares). In order to classify each available point, a distance metric to calculate the maximum neighbor distance for each point first has to be defined. In [4], the distance used is a Euclidean distance based on the range- and velocity-bins in the RDIs. A point is then classified as

1. a core point, if it has at least $N - 1$ neighbors (points within the maximum neighbor distance)
2. an edge point, if it has less than $N - 1$ neighbors, but at least one edge point as neighbor
3. noise otherwise.

The performance of this processing chain, consisting of OS-CFAR detection and DBSCAN clustering, depends on the correct setting of the hyperparameters of

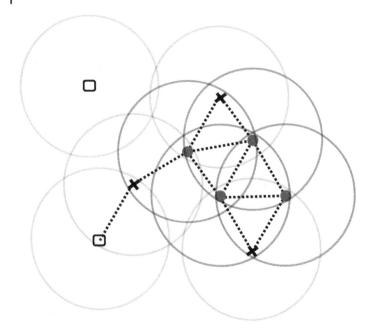

Figure 6.11 Visualization of core (circles), edge (crosses), and noise-points (squares) in DBSCAN for $N = 4$.

the single processing blocks. Even if well adjusted, optimal hyperparameters may still vary with the input data, potentially resulting in ghost targets or missed detections. In the output of the signal processing chain in Figure 6.11, some of these issues are highlighted with boxes. Figure 6.12 shows a detection chain for the same segmentation task but with a neural network replacing the traditional signal processing blocks.

The internal structure of the neural network is shown in Figure 6.13. It has a U-Net-like [29] convolutional autoencoder structure with skip connections

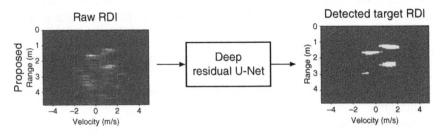

Figure 6.12 Deep learning processing chain for target detection.

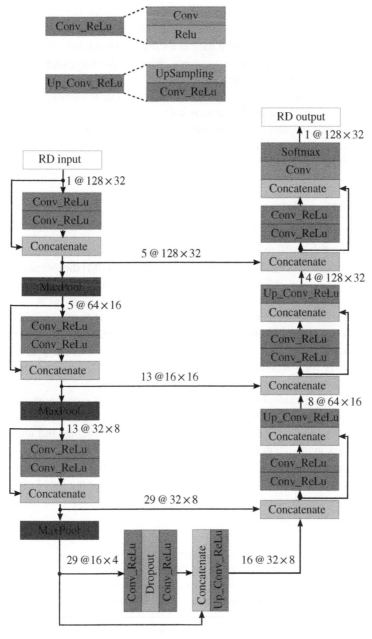

Figure 6.13 Deep learning processing chain for target segmentation.

between the encoder part of the network and the decoder. One idea of these skip connections is to better preserve spatial information of the clusters in the RDIs to help the segmentation task. The training dataset for the deep residual U-Net used in [4] consists of real one-target measurements, where radar recordings were taken with one human target walking around in an office space and multitarget measurements with the corresponding ground-truth data. The multitarget measurements and the multitarget label data were created by superimposing the raw RDIs of different real one-target measurements with some restrictions to avoid overlaps of clusters. The loss function used to learn the segmentation task is a weighted combination of focal loss and hinge loss:

$$HL(p) = 1 - y(2p - 1)$$
$$FL(p_t) = (1 - p_t)^\gamma \log(p_t) \qquad\qquad (6.14)$$
$$L(p_t) = \alpha[FL(p_t) + \mu HL(p)]$$

Here, HL, FL, and L describe the hinge loss, the focal loss, and the combined loss term, respectively, with $y \in \{\pm 1\}$ describing the class label, $p \in [0,1]$ the estimated probability of $y = 1$, and $p_t \in [0,1]$ the probability of the correct classification of each pixel.

Table 6.2 shows the performance of the deep learning-based U-Net approaches compared to the traditional method with OS-CFAR and DBSCAN clustering in terms of F1-score on a test set. Both different network sizes show a large performance gain over the traditional method, with the deeper network achieving slightly higher accuracy.

Figure 6.14 shows an example for a four people human target scene, with the raw radar data after moving target indicator (MTI) and FFTs in Figure 6.14a, and the segmented versions with OS-CFAR plus DBSCAN in Figure 6.14b, and with the neural network approach in Figure 6.14c.

While the neural network manages to correctly identify and segment the four clusters belonging to the four human targets, the signal processing-based processing chain has some missed detections and false alarms, marked with boxes in Figure 6.14b.

Table 6.2 Comparison of the detection performance of the traditional pipeline with the proposed U-Net architecture with a depth of 3 and depth of 5 for people counting.

Approach	Description	F-score	Model size
Traditional	OS-CFAR with DBSCAN	0.71	—
Proposed U-Net depth 3	Proposed loss	0.89	616 kB
Proposed U-Net depth 5	Proposed loss	0.91	2.8 MB

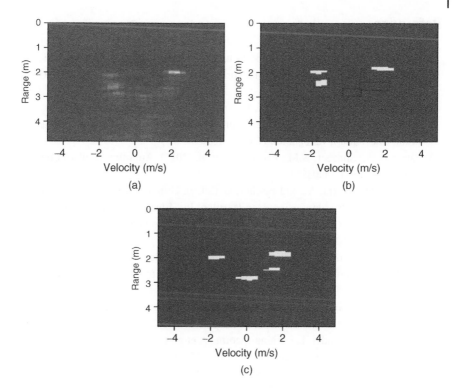

Figure 6.14 (a) Raw RDI image with four human targets, (b) processed RDI using the traditional approach wherein one target is split and two targets are occluded, (c) processed RDI using the neural network approach wherein all targets are detected accurately.

6.6 Questions to the Reader

- What are some different methods to inject domain knowledge into deep learning problems?
- Why do you sometimes want to restrict the expressiveness of the network by injecting domain knowledge?
- How can the dictionary be represented in the learned ISTA for DoA estimation with radar?
- After reading about algorithm unrolling in the sense of learned ISTA for radar, how do you think that the network might achieve superior results to the base algorithm?
- When does it make sense to use the focal loss and deformable convolutions?
- Given what you have read in the last section, what may allow the neural network to outperform traditional signal processing in radar target segmentation?

References

1 Krizhevsky, A., Sutskever, I., and Hinton, G.E. (2017). ImageNet classification with deep convolutional neural networks. *Communications of the ACM* 60 (6): 84–90.

2 Redmon, J., Divvala, S., Girshick, R., and Farhadi, A. (2016). You only look once: unified, real-time object detection. *2016 IEEE Conference on Computer Vision and Pattern Recognition (CVPR)*. IEEE.

3 Vaswani, A., Shazeer, N.M., Parmar, N. et al. (2017). Attention is all you need. *ArXiv*, vol. abs/1706.03762.

4 Stephan, M., Santra, A., and Fischer, G. (2020). Human target detection and localization with radars using deep learning. In: *Advances in Intelligent Systems and Computing* (ed. M. Arif Wani, Taghi M. Khoshgoftaar, and Vasile Palade), 173–197. Singapore: Springer.

5 Stephan, M., Hazra, S., Santra, A. et al. (2021). People counting solution using an FMCW radar with knowledge distillation from camera data. *2021 IEEE Sensors*.

6 Hazra, S. and Santra, A. (2018). Robust gesture recognition using millimetric-wave radar system. *IEEE Sensors Letters* 2 (4): 1–4.

7 Stephan, M., Servadei, L., Arjona-Medina, J. et al. (2022). Scene-adaptive radar tracking with deep reinforcement learning. *Machine Learning with Applications* 100284. https://www.sciencedirect.com/science/article/pii/S2666827022000196.

8 Gregor, K. and LeCun, Y. (2010). Learning fast approximations of sparse coding. *ICML*.

9 Zhang, K., Gool, L.V., and Timofte, R. (2020). Deep unfolding network for image super-resolution. *2020 IEEE/CVF Conference on Computer Vision and Pattern Recognition (CVPR)*, pp. 3214–3223.

10 Zhang, Z., Liu, Y., Liu, J. et al. (2021). AMP-Net: Denoising-based deep unfolding for compressive image sensing. *IEEE Transactions on Image Processing* 30: 1487–1500.

11 Kittenplon, Y., Eldar, Y.C., and Raviv, D. (2021). FlowStep3D: Model unrolling for self-supervised scene flow estimation. *2021 IEEE/CVF Conference on Computer Vision and Pattern Recognition (CVPR)*, pp. 4112–4121.

12 He, H., Wen, C.-K., Jin, S., and Li, G.Y. (2020). Model-driven deep learning for MIMO detection. *IEEE Transactions on Signal Processing* 68: 1702–1715.

13 Monga, V., Li, Y., and Eldar, Y.C. (2021). Algorithm unrolling: interpretable, efficient deep learning for signal and image processing. *IEEE Signal Processing Magazine* 38 (2): 18–44.

14 Natarajan, B.K. (1995). Sparse approximate solutions to linear systems. *SIAM Journal on Computing* 24 (2): 227–234 https://doi.org/10.1137/s0097539792240406.

15 Candès, E.J., Romberg, J.K., and Tao, T. (2005). Stable signal recovery from incomplete and inaccurate measurements. *Communications on Pure and Applied Mathematics* 59: 1207–1223.

16 Ramírez, C., Kreinovich, V., and Argáez, M. (2013). Why l1 is a good approximation to l0: a geometric explanation.

17 Fu, R., Huang, T., Liu, Y., and Eldar, Y.C. (2019). Compressed LISTA exploiting Toeplitz structure. *2019 IEEE Radar Conference (RadarConf)*. IEEE.

18 Chipengo, U., Sligar, A.P., Canta, S.M. et al. (2021). High fidelity physics simulation-based convolutional neural network for automotive radar target classification using micro-Doppler. *IEEE Access* 9: 82597–82617.

19 Chen, V., Li, F., Ho, S.-S., and Wechsler, H. (2006). Micro-Doppler effect in radar: phenomenon, model, and simulation study. *IEEE Transactions on Aerospace and Electronic Systems* 42 (1): 2–21.

20 Erol, B., Gurbuz, S.Z., and Amin, M.G. (2020). Motion classification using kinematically sifted ACGAN-synthesized radar micro-Doppler signatures. *IEEE Transactions on Aerospace and Electronic Systems* 56 (4): 3197–3213.

21 Stephan, M., Stadelmayer, T., Santra, A. et al. (2021). Radar image reconstruction from raw ADC data using parametricvariational autoencoder with domain adaptation. *2020 25th International Conference on Pattern Recognition (ICPR)*. IEEE.

22 Shankar, Y., Hazra, S., and Santra, A. (2019). Radar-based non-intrusive fall motion recognition using deformable convolutional neural network. *2019 18th IEEE International Conference On Machine Learning And Applications (ICMLA)*. IEEE.

23 Lin, T.-Y., Goyal, P., Girshick, R. et al. (2017). Focal loss for dense object detection. *2017 IEEE International Conference on Computer Vision (ICCV)*, pp. 2999–3007.

24 Chopra, S., Hadsell, R., and LeCun, Y. (2005). Learning a similarity metric discriminatively, with application to face verification. *2005 IEEE Computer Society Conference on Computer Vision and Pattern Recognition (CVPR'05)*, Volume 1, pp. 539–546.

25 Hochreiter, S. and Schmidhuber, J. (1997). Long short-term memory. *Neural Computation* 9 (8): 1735–1780.

26 Springenberg, J.T., Dosovitskiy, A., Brox, T., and Riedmiller, M.A. (2015). Striving for simplicity: the all convolutional net. *CoRR*, vol. abs/1412.6806.

27 Will, C., Vaishnav, P., Chakraborty, A., and Santra, A. (2019). Human target detection, tracking, and classification using 24-GHz FMCW radar. *IEEE Sensors Journal* 19 (17): 7283–7299.

28 Xu, D. and Tian, Y. (2015). A comprehensive survey of clustering algorithms. *Annals of Data Science* 2 (2): 165–193.

29 Ronneberger, O., Fischer, P., and Brox, T. (2015). U-Net: Convolutional networks for biomedical image segmentation. *International Conference on Medical image computing and computer-assisted intervention*. Springer, pp. 234–241.

7

Domain Adaptation

After reading this chapter, you should be able to

- Understand the basic concept of domain adaptation.
- Have an overview of different types and methods of domain adaptation.
- Understand how domain adaptation can help to improve real-world radar applications.

7.1 Introduction

In this chapter, we review how domain adaptation (DA), a field of machine learning (ML) where an algorithm learns from a source data distribution to perform on a related target data distribution, works on radar signal processing. To this end, we first define DA and position it in the parent category of transfer learning (TL). Successively, we explore different subcategories of DA and common data shifts between source and target domain. We then present an overview of DA algorithms as well as the background work of DA in radar signal processing. Finally, we present an application of Domain Adaptation as well as the conclusion of the chapter.

7.2 Transfer Learning and Domain Adaptation

Domain adaptation is a subcategory of TL. TL is a research problem in ML that focuses on storing knowledge gained while solving one problem and applying it to a different but related problem [1]. As shown in Figure 7.1, while this definition applies to differences in the target domain's feature space and source feature space,

Methods and Techniques in Deep Learning: Advancements in mmWave Radar Solutions, First Edition.
Avik Santra, Souvik Hazra, Lorenzo Servadei, Thomas Stadelmayer, Michael Stephan, and Anand Dubey.
© 2023 The Institute of Electrical and Electronics Engineers, Inc. Published 2023 by John Wiley & Sons, Inc.

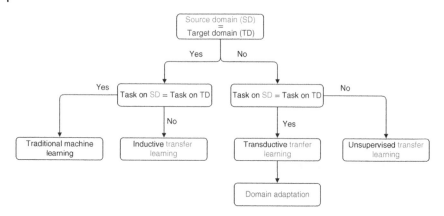

Figure 7.1 Transfer learning and domain adaptation.

DA assumes that the task in source and target domain is the same. Even if, there is a change in distributions.

In order to thoroughly define domain adaptation, we focus on its formalization and the relationship with TL, similarly as done in [2].

As a start, we consider a d-dimensional feature space $X \subset R^d$ in a domain D.

The marginal probability of the random variable X is identified as $P(X)$ and a task T is identified by a label space Y, another random variable, and the conditional probability $P(Y|X)$.

In order to learn in a supervised fashion, we consider two sample sets of size n, namely $X = \{x_1, x_2, \ldots, x_n\}$ from the X and $Y = \{y_1, y_2, \ldots, y_n\}$ from the Y space. Using the feature-label pairs $\{x_i, y_i\}$, we can learn the conditional distribution $P(Y|X)$.

Once the number of domains and tasks increases, i.e., two domains and two different tasks, things become more challenging.

We then call the first domain, the source domain, and identify it by $D^S = \{X^S, P(X)^S\}$. The second domain is called target domain and is identified by $D^T = \{X^T, P(X)^T\}$. Accordingly, the two tasks, namely source and target tasks, are denoted as $T^S = \{Y^S, P(Y^S|X^S)\}$ and $T^T = \{Y^T, P(Y^T|X^T)\}$.

In case $D^S = D^T$ and $T^S = T^T$, we have a standard ML framework where D^S is the training set, and D^T corresponds to the test set. If one of the two conditions is missing, we could have a poor performance by using a model trained on the source domain, for the target task. In fact, standard learning theory and model guarantees do not hold if the two domains are different (i.e., $D^S \neq D^T$) or the two tasks for which the model is trained to differ (i.e., $T^S \neq T^T$).

Nevertheless, in case we have some relatedness between the two domains, we could exploit the source domain and its task, namely D^S, T^S, to learn $P(Y^T, X^T)$. This process is defined as TL.

To this end, homogeneous TL is the case where $X^T = X^S$, but the marginal probabilities are different, such as $P(X^T) \neq P(X^S)$, due to domain shift. Heterogeneous TL instead takes place if the source and target data present significant differences either in terms of representation or even structure, such that $X^T \neq X^S$.

Considering those cases, TL is generally categorized by the relation between source and target domain, as well as the corresponding tasks.

These are inductive TL, transductive TL, and unsupervised TL.

- *Inductive TL* is the case where the target task is related to the source task, with no further assumption on the relatedness of the source and target domains. In order to apply TL in this case, at least a small set of labels needs to be available, to induce the model on the target task.
- *Transductive TL* is the category that concerns the same source and target task. In this case, if source and target domain differ, we have domain adaptation.
- *Unsupervised TL* identifies instead the category where both the domains and the tasks are different but are somehow related. Typically, labels are not provided for any of the domains, and the scope is extracting useful characteristics in the source domain, which can be used for approaching a task in the target domain (e.g., clustering, density estimation).

As presented in these categories, DA methods belong to transductive TL, where $T^T = T^S$. This entails that both the set of labels and the conditional distributions are assumed to be shared between the two domains, i.e., $Y^S = Y^T$ and $P(Y|X^T) = P(Y|X^S)$. While this is formally correct, often in real life, the second assumption is not fulfilled. For this reason, the first condition is generally enough to define DA.

In further cases, domain adaptation can be found where labels are provided for both source and target domain (i.e., *supervised*), only on the source domain (i.e., *unsupervised*), or partially to the target and fully to the source domain (i.e., *semi-supervised*).

7.3 Categories of Domain Adaptation

Domain adaptation is usually divided into three categories, as shown in Figure 7.2, namely:

- *Unsupervised domain adaptation*, where the learning sample contains a set of labeled source examples and a set of unlabeled target examples.
- *Semisupervised domain adaptation* we also consider a 'small' set of labeled target examples.
- *Supervised domain adaptation* where all the examples considered are labeled.

Supervised domain adaptation   Semi-supervised domain adaptation   Unsupervised domain adaptation

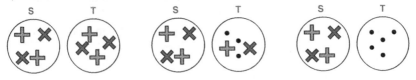

Figure 7.2 Categories of domain adaptation.

7.3.1 Common Data Shifts

Data shifts are common mismatches that can be found on different datasets, dataset, such as for the training dataset and data in the inference process or source domain distribution and target domain distribution.

Overcoming this shift is a subject of research, and traditional ML algorithm performs poorly than the original and shifted distribution.

7.3.1.1 Prior Shift

In the prior shift, while the probabilities for labels differ (i.e., $P^S(Y) \neq P^T(Y)$), the conditional probabilities stay the same, $P^S(Y|X) = P^T(Y|X)$.

This can happen, for example, while predicting the sales of a shopping mall given the income of the customers, while it is in a poorer or wealthier neighborhood.

7.3.1.2 Covariate Shift

Covariate shift happens instead when the distribution of the input data (or independent variables) has a shift. In this case, $P^S(Y|X) = P^T(Y|X)$ but $P^S(X) \neq P^T(X)$. A case for it might be the risk of diseases per age two cities. While the conditional on the risk remains the same, the number of older people in one city can be higher than the other one.

7.3.1.3 Concept Shift

Finally, concept shift refers to the case where conditional probabilities are changed, but not the marginals: $P^S(Y|X) \neq P^T(Y|X)$ but $P^S(X) = P^T(X)$. An example of it is the difference in housing prices given the number the same number of rooms in two different datasets.

7.3.2 Methods of Domain Adaptation

Deep domain adaptation utilizes deep neural network architectures to enhance the performance of DA algorithms. To this end, similarly as in other ML categories (e.g., reinforcement learning), using deep networks enhances the approximation capability of the methods and leads to surpassing the state of the art. In the particular case of domain adaptation, the capacity of deep networks for extracting more expressive features is a clear enhancement toward reaching transferable and readaptable distributions from the source to the target domain.

Methods of Domain Adaptation can be further subcategorized as follows:

- Discrepancy-based
- Adversarial-based
- Reconstruction-based

7.3.2.1 Discrepancy-based Domain Adaptation

The category of discrepancy-based methods encourages the fine-tuning of deep network models in order to diminish the shift between the source and target domain. Therefore, different approaches can be used to perform the fine-tuning. In the next paragraphs, we are going to review a selection of them, similarly to the classification done in [3].

Using class label information, knowledge can be transferred between source and target domain. For supervised DA, soft labels and metric learning can be used for enhancing the transfer process to the target domain. For unsupervised DA cases, instead, pseudolabels and attribute representation techniques can substitute the missing labels in the fine-tuning process, as shown in [4, 5].

Other methods use statistics to reduce the shift between domains: a popular method for reducing the distribution shift is the maximum mean discrepancy. There it is defined by the idea of representing distances between distributions as distances between mean embeddings of features [6]. This distance is minimized in order to bridge the two domains feature representation.

Another approach to improve DA using discrepancy-based techniques focuses on improving network architectures for a better capturing and transferring of features. The most popular method used in this category is the adaptive batch normalization [7]. This allows the training to be optimized to locate features which can be easily transferred to a following task. Other methods make use of weight regularization terms between layers of the source and target networks to punish large deviations in layer weights and therefore force the networks to extract similar features. This weight regularization loss term is typically based on some layerwise distance between the parameters of the source network, after training with data and labels from the source domain to the parameters of the target network.

$$L_{\mathrm{DA}} = \sum_{i \in K} \lambda_i f(\Theta_i^{\mathrm{T}}, \Theta_i^{\mathrm{S}}) \tag{7.1}$$

Equation (7.1) shows a general form of such a loss term which can be added to the overall loss function. Here, K is some set of layers in the networks, f describes some distance function, λ_i are some weighting factors, Θ_i^{T} are the current parameters of the target network, and Θ_i^{S} are the parameters of the source networks after training.

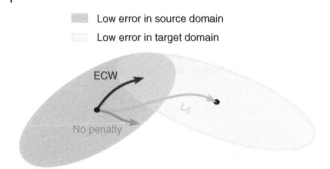

Figure 7.3 Elastic weight consolidation helps in finding a common low-error region for both domains. A simple L_2-error may restrict the network weights too much, so that the performance suffers and a low-error region in the target domain is not reached. Using no weight penalty may result in forgetting the source domain task, which is often not desired.

While a simple normalized distance function for the weights and biases in each layer, as in [8], has been used, more sophisticated approaches, such as elastic weight consolidation [9], have shown promising results. The main idea of elastic weight consolidation is to use a weight regularization term which punishes deviations for parameters that were important for the predictions in the source domain harsher than for those that were less important. This can help in better preventing catastrophic forgetting on the source domain task, as visualized in Figure 7.3. The necessary information is contained in the posterior probability $P(\Theta^T|D^S)$, which describes the influence of the source domain data on the network parameters Θ^T. As calculating this posterior distribution is intractable, the diagonal elements of the Fisher information in matrix form are used instead, as shown in Eq. (7.2).

$$L_{DA} = \sum_i \frac{\lambda}{2} F_i(\Theta_i^T, \Theta_i^S)^2 \qquad (7.2)$$

The Fisher information can be used as an approximation to measure the importance of the parameters for the source-domain task. It can easily be calculated directly from the parameter gradients near the optimum for those trained on the source domain.

Finally, a set of methods is used in the literature which exploits geometrical properties of source and target domains to reduce the domain shift, as shown in [10].

7.3.2.2 Adversarial-based Domain Adaptation

Another major category of DA methods is called adversarial-based deep DA. This concentrates on adversarial training for a better task transferability between source and target domains.

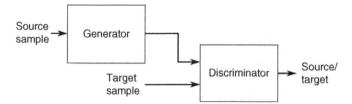

Figure 7.4 Illustration of a GAN architecture for domain adaptation. The discriminator tries to classify from which domain its input came from. The generator tries to fool the discriminator by mapping source-domain samples to the target domain.

In this case, an adversarial objective, as seen in Eq. (7.3), is used to minimize the distance between the source and target distribution.

$$\min_D \max_G E_{x \sim p_{\text{data}}(x)}[\log(D(x))] + E_{z \sim p_z(z)}[\log(1 - D(G(z)))] \tag{7.3}$$

The core idea of the loss equation for generative adversarial networks is the competition between the generator network G and the discriminator network D. Given the generator output and some real label, the discriminator tries to minimize the objective in Eq. (7.3) – it tries to detect whether or not its input comes from the generator. The generator instead tries to maximize the same objective, so its goal is to fool the generator into classifying the generator outputs as real labels.

For the domain adaptation task, the input to the generator could be a sample from the source distribution, whereas the label input to the discriminator could be a sample from the target distribution, as seen in Figure 7.4. On successful training, the generator then transforms samples from the source domain to target domain samples. A common application would be the transformation of synthetically generated to data samples that are indistinguishable from real ones.

Another option uses another encoder network instead of the generator, according to [11], as seen in Figure 7.5. Here, input samples from either the source domain or the target domain are fed into the encoder which produces some feature vector as output. This feature vector is then further processed in a classifier network to predict the correct label, whereas a discriminator or domain classifier network tries to predict whether the produced feature vector stems from the source or the target domain. The overall loss function is then a weighted combination of the discriminator loss and the class label loss. As indicated by the circled and crossed arrows in Figure 7.5, the sign of the gradients of the discriminator loss function with respect to the encoder parameters Θ_e is reversed. By minimizing the overall loss function, the encoder will then partly try to maximize the loss of the discriminator with respect to the encoder parameters. In other words, it will try to fool the discriminator in a generative adversarial network (GAN)-fashion and therefore produce similar feature vectors for source and target domain samples.

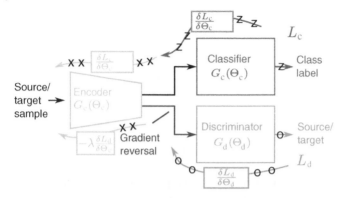

Figure 7.5 Adversarial DA architecture without the generator. Source: Adapted from Ganin and Lempitsky [11]. The discriminator has to detect whether the encoded input feature vector stems from the source or the target domain. The same classifier works on both domains.

7.3.2.3 Reconstruction-based Domain Adaptation

The third main category of DA approaches is called reconstruction-based. Reconstructing the initial data, in fact, we can preserve information and be able, at the same time, to capture essential features for each of the domains, as presented in [12].

Two main approaches used are encoder–decoder reconstruction, which uses the traditional autoencoders or similar approaches as sketched in Figure 7.6, or the adversarial reconstruction, utilizing GAN models or slight variations of it (e.g., Cycle GAN).

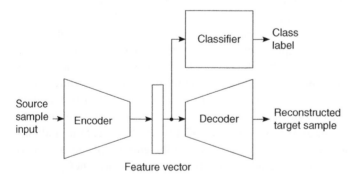

Figure 7.6 Sketch for a reconstruction-based domain adaptation using an autoencoder. The decoder tries to reconstruct the target sample, given an encoded source sample.

7.4 Domain Adaptation in Radar Processing

While the use of deep learning methods for radar applications has become more popular in recent times, the lack of usable public datasets and training data remains an issue. Using available training data from a different source domain can potentially alleviate that problem. For radar processing tasks, one can typically further split the domain adaptation approaches by the choice of the source domain. The source domain data can either come from

1. a different type of sensor
2. or another radar sensor with different settings.

In the following subsections, we will look at some different examples from literature for both of these cases.

7.4.1 Domain Adaptation with a Different Sensor Type

For the first case, a common choice for the source domain sensor is either light detection and ranging (LiDAR) or camera, both of which are often used in similar tasks or in combination with each other, as is often the case in autonomous driving. Especially camera data can be very attractive due to the large amount of publicly available training data for all kinds of different applications.

In [13], LiDAR, camera, or radar data are used as source-domain data, and radar or LiDAR as the target domain. Here, two autoencoders, one for the source domain and one for the target domain, try to reconstruct their respective input data. A maximum mean discrepancy loss function, given by Eq. (7.4), between the embeddings of the autoencoders is used during their training to ensure the extraction of similar features.

$$L_{\mathrm{MMD}}(\hat{X}^{\mathrm{S}}, \hat{X}^{\mathrm{T}}) = \left\| \frac{1}{N^{\mathrm{S}}} \sum_{i=1}^{N^{\mathrm{S}}} \hat{X}_i^{\mathrm{S}} - \frac{1}{N^{\mathrm{T}}} \sum_{i=1}^{N^{\mathrm{T}}} \hat{X}_i^{\mathrm{T}} \right\|^2 \tag{7.4}$$

In Eq. (7.4), N^{T} and N^{S} are the number of samples of the target domain and the source domain, respectively. $\hat{X}^{\mathrm{S}}$ and $\hat{X}^{\mathrm{T}}$ describe the embeddings of the respective domains. After training both autoencoders in this manner, they achieve an embedding space that is less variant to the input feature domain. In the next step, they freeze the encoder weights, take the source encoder, and append a classifier network to it. After supervised training of the classifier network they take, they cascade the target encoder network and the source classifier network to do another training round on the target domain data. The training procedure is visually described in Figure 7.7.

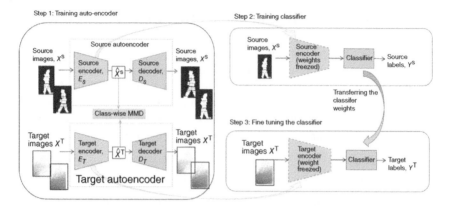

Figure 7.7 Autoencoder and classifier training method according to Rahman et al. Source: Rahman et al. (2021) [13]. IEEE.

With experiments, they demonstrate the improved performance compared to training only on the radar or LiDAR target domain data.

The other possibility is making use of data recorded with radar but under different settings. These different settings can include different hardware (different number of antennas/antenna configuration/hardware noise…), varying waveform settings (number of samples, chirps, bandwidth, pulse repetition time, observation interval, …), or changes in the environment in which the training data is recorded (indoors, outdoors, open spaces, spaces with a lot of potential reflectors, synthetic data, …). Due to the limited amount of labeled radar data available, in [8], domain adaptation, with synthetic-generated radar-data as source domain for the task of human segmentation in radar images is used. While the synthetic data here stems from simple point-target simulations and does not represent real human targets, it can still be used to learn the extraction of high-level information like the direction of arrival of targets. Here, they use a discrepancy-based method, by first training a network with the synthetic data, and then adding another loss term to the training with real data, to ensure that similar features are extracted.

$$L_{DA} = \sum_{i \in K} \frac{\left\| w_i - w_{i0} \right\|^2 + \left\| b_i - b_{i0} \right\|^2}{\left\| w_{i0} \right\|^2 + \left\| b_{i0} \right\|^2} \qquad (7.5)$$

Equation 7.5 describes said additional loss term L_{DA}. Here, w and b are the target network weights and biases, while w_0 and b_0 are the source network weights and biases. The equation then describes the sum over normalized difference between weights and biases for the set of convolutional layers K. It is demonstrated in the study that adding this loss term keeps the target network from deviating too far from the source network without affecting the performance on the target domain.

7.4.2 Domain Adaptation with Different Radar Settings

7.4.2.1 Introduction

Radar, a well-established technology for several industrial areas, has recently gained attention for other commercial applications, such as human monitoring, presence detection, or gesture sensing [14–16] due to the production of small and compact radar sensors [17]. Here, radar offers some advantages in comparison with computer vision approaches, such as good performance under poor-lighting conditions or privacy protection (due to the difficulty in identifying individuals from radar images).

Similarly as in the realm of computer vision, the use of radar often comes hand in hand with ML techniques, including deep learning, to overcome the burden of handcrafted feature engineering [18]. Due to the variety of system design parameters at hand, such as modulation techniques or bandwidth, these ML algorithms are required to generalize well under different radar setups. This need for interdomain generalization is common to several ML problems, and it has been studied in recent years under the paradigm of domain adaptation [19, 20].

Considered as a special case of TL, domain adaptation involves modifying an ML estimator that can be trained with enough data from a *source domain*, so that its performance increases when evaluated with data originating from a different *target domain*. Unlike other TL approaches, here, the mismatch between source and target domains lies merely in a distinct probability measure over data rather than in different input or output spaces [20].

The reasons for domain adaptation are usually related to insufficient or incomplete data in the target domain, which can be overcome with the help of data from the source domain. In ML classification, the missing information is often the labels; this case is referred to as *unsupervised* domain adaptation. If target data are labeled, we can apply *supervised* domain adaptation techniques instead [19].

Both supervised and unsupervised domain adaptation methods have already been investigated in the Radar–ML community to overcome several problems, including individual patient differences [21], aspect angle variations [22], synthetic-to-real adaptation [23] or environmental differences [8]. In the case of cross-configuration adaptation, Khodabakhshandeh et al. [24] use supervised techniques such as few-shot adversarial domain adaptation (FADA) [25] or DSNE [26] to adapt their trained human activity classifier to new FCMW radar setups using few data.

We build on the work in [24] by applying margin disparity discrepancy (MDD) [27]. In that way, we confirm that this unsupervised technique, which delivers state-of-the-art results for computer vision datasets, also works for radar data and thus enables cross-configuration radar-based human activity classification based on unlabeled data.

7.4.2.2 Problem Statement

Radar-ML classification deals with the evaluation of a group of radar features $x \in \mathcal{X}$ obtained from a target to find the underlying class $y \in \mathcal{Y}$ that best describes some property of the said target. The input space $\mathcal{X} \subset \mathbb{R}^m$ is characterized by a dimension m that depends on the radar technology and preprocessing steps, while the label space is defined as $\mathcal{Y} = \{1, \ldots, k\}$, with k being the number of classes.

In order to achieve this classification, one has first to find a classifier h that maps x into y. The ML approach assumes a sufficiently large amount of data available conveying information both about the inputs and the class so that it can be used to train h among a restricted hypothesis class $\mathcal{H}$. This dataset consists of a sequence of pairs of features and labels, i.e., $\{(x_i, y_i)\}_{i=1}^{n}$, that have been previously sampled from a certain domain D, defined to be

$$D = (\mathcal{X}, \mathcal{Y}, p_D), \tag{7.6}$$

with an associated probability measure p_D over $\mathcal{X} \times \mathcal{Y}$. Here and hereafter, we write $\mathbf{x}$ and $\mathbf{y}$ in their upright form whenever we refer to the random variables related to p_D, and not to its realizations.

By choosing an objective loss function

$$\ell : \mathcal{H} \times \mathcal{X} \times \mathcal{Y} \to \mathbb{R}_{0+} \tag{7.7}$$

and minimizing it over the hypothesis class $\mathcal{H}$ with a suitable optimization method, we can train an h that performs well for the available data. The performance of h can thus be measured by the *risk* associated with the loss for a domain D. This risk $\mathcal{L}_D$ represents the expected value of the loss of h for p_D:

$$\mathcal{L}_D(h) = \mathbb{E}_D \ell(h, (\mathbf{x}, \mathbf{y})). \tag{7.8}$$

By assuming the indicator function $\mathbb{1}_{h(x) \neq y}$ to be the loss, we obtain the *0-1 error* $\mathrm{err}_D(h) \triangleq \mathbb{E}_D \mathbb{1}_{h(x) \neq y}$. In practice, we do not have access to p_D, so we resort to its empirical approximation $\mathcal{L}_{\hat{D}}(h) \triangleq \sum_{i=1}^{n} \ell(h, (x_i, y_i)) / n$ for a dataset $\hat{D}$ with n samples drawn from D.

If generalization is achieved, h will also behave well for unseen data as long as it is drawn from the same domain. Unfortunately, this assumption cannot always be guaranteed. It is often the case that training data have been drawn from a *source domain* S, but we would like to leverage the trained classifier for a different *target domain* $\mathcal{T}$. Depending on how dissimilar S and $\mathcal{T}$ are, the performance of the trained classifier can degrade significantly. In our specific problem, this *domain shift* is given by the choice of different FCMW settings and presents an additional challenge in the lack of the labels for the training data from $\mathcal{T}$. The absence of labeled target data makes it necessary to apply *unsupervised* domain adaptation. We explore this possibility by using MDD [27].

7.4.2.3 MDD

In order to use MDD, we assume a hypothesis class induced by a space $\mathcal{F}$ of *scoring functions* $f : \mathcal{X} \mapsto \mathbb{R}^k$. We also introduce the shorthand $f_y(x)$ to refer to the yth component of $f(x)$. The hypothesis class is given by

$$\mathcal{H} \triangleq \left\{ h_f : x \mapsto \arg\max_{y \in \mathcal{Y}} f_y(x) \mid f \in \mathcal{F} \right\} \tag{7.9}$$

MDD has been developed by Zhang et al. [27] as a practical algorithm based on the concept of discrepancy distance by Mansour et al. [28]. For that, they define the *margin error* $\text{err}_D^{(\rho)}(f)$ as

$$\text{err}_D^{(\rho)}(f) \triangleq \mathbb{E}_D \Phi^{(\rho)} \circ \phi_f(x, y) \tag{7.10}$$

$$\phi_f(x, y) \triangleq \frac{1}{2} \left(f_y(x) - \max_{y' \neq y} f_{y'}(x) \right) \tag{7.11}$$

$$\Phi^{(\rho)}(x) \triangleq \begin{cases} 0 & \rho \leq x \\ 1 - x/\rho & 0 \leq x \leq \rho \\ 1 & x \leq 0 \end{cases} \tag{7.12}$$

and the true and empirical *margin disparity* between two scoring functions f' and f as follows:

$$\text{disp}_D^{(\rho)}(f', f) \triangleq \mathbb{E}_D \Phi^{(\rho)} \circ \phi_{f'}(x, h_f(x)) \tag{7.13}$$

$$\text{disp}_{\hat{D}}^{(\rho)}(f', f) \triangleq \frac{1}{n} \sum_{i=1}^{n} \Phi^{(\rho)} \circ \phi_{f'}(x_i, h_f(x_i)) \tag{7.14}$$

to finally formulate the following minimax optimization problem:

$$\min_{f \in \mathcal{F}} \text{err}_{\hat{S}}^{(\rho)}(f) + d_{f,\mathcal{F}}^{(\rho)}\left(\hat{S}, \hat{\mathcal{T}}\right)$$
$$d_{f,\mathcal{F}}^{(\rho)}(S, \mathcal{T}) \triangleq \sup_{f' \in \mathcal{F}} \left(\text{disp}_{\mathcal{T}}^{(\rho)}(f', f) - \text{disp}_S^{(\rho)}(f', f) \right) \tag{7.15}$$

Following the principles of unsupervised domain adaptation, the MDD term $d_{f,\mathcal{F}}^{(\rho)}$ does not make use of any labels y_i. Furthermore, the solution to (7.15) minimizes the 0–1 error of h_f in the target domain, as it is [27] proved with the following theoretical bound:

$$\text{err}_{\mathcal{T}}(h_f) \leq \text{err}_S^{(\rho)}(f) + d_{f,\mathcal{F}}^{(\rho)}(S, \mathcal{T}) + \lambda \tag{7.16}$$

where λ is the ideal combined margin loss:

$$\lambda = \min_{f^* \in \mathcal{F}} \left\{ \text{err}_S^{(\rho)}(f^*) + \text{err}_{\mathcal{T}}^{(\rho)}(f^*) \right\}. \tag{7.17}$$

The bound in (7.16) can also be expressed in terms of empirical measures rather than true probability measures by the addition of Rademacher complexity terms [29].

Despite the interesting properties of MDD, Zhang et al. [27] ultimately resort to the cross-entropy loss instead of the margin loss in order to avoid vanishing and exploding gradients during training. Prior to that, they map $f(x)$ to the k-simplex via the softmax function σ, as it is customary in deep learning, where the elements of $\sigma(z)$ are given by

$$\sigma_j(z) \triangleq \frac{\exp z_j}{\sum_{i=1}^{k} \exp z_i}, \quad \text{for } j = 1, \dots, k \tag{7.18}$$

The composition of the cross-entropy loss with the softmax yields the least square error (LSE) loss $\mathcal{L}_f$:

$$\begin{aligned} \mathcal{L}_f(x, y) &\triangleq -\log \sigma_y(f(x)) \\ &= \log \sum_{y' \in \mathcal{Y}} \exp\left(f_{y'}(x) - f_y(x)\right) \end{aligned} \tag{7.19}$$

[27] propose to use $\mathcal{L}_f$ instead of $\Phi^{(\rho)} \circ \phi_f$ for $\mathrm{err}_{\hat{S}}^{(\rho)}$ and $\mathrm{disp}_{\hat{S}}^{(\rho)}$ in (7.15). As for $\mathrm{disp}_{\hat{T}}^{(\rho)}$, they use the adversarial loss $\tilde{\mathcal{L}}_f$ proposed by [30], i.e.,

$$\tilde{\mathcal{L}}_f(x, y) \triangleq \log\left(1 - \sigma_y(f(x))\right) \tag{7.20}$$

so that their MDD ultimately becomes

$$\begin{aligned} \tilde{d}_{f,\psi,\mathcal{F}}^{(\gamma)}\left(\hat{S}, \hat{T}\right) &\triangleq \max_{f' \in \mathcal{F}} \mathbb{E}_{x^t \sim \hat{T}} \tilde{\mathcal{L}}_{f'}\left(\psi\left(x^t\right), h_f\left(\psi\left(x^t\right)\right)\right) \\ &\quad - \gamma \mathbb{E}_{x^s \sim \hat{S}} \mathcal{L}_{f'}\left(\psi\left(x^s\right), h_f\left(\psi\left(x^s\right)\right)\right) \end{aligned} \tag{7.21}$$

for a *margin factor* $\gamma > 0$ and a feature extractor ψ that levels the min-player to the max-player [27] (A concrete example of ψ is given in (7.32)). It is explained that this is equivalent to the use of the margin loss with a margin $\rho = \log \gamma$ and that the problem is still solved for $S = \mathcal{T}$ [27].

In addition to the results in [27], we observe that the use of the recently proposed soft-margin softmax $\sigma^{(\rho)}$ [31] instead of σ in (7.19) provides an upper bound for $\mathrm{err}_S^{(\rho)}$. The entries of $\sigma^{(\rho)}(z)$ are defined as follows:

$$\sigma_j^{(\rho)}(z) \triangleq \frac{\exp\left(z_j - \rho\right)}{\exp\left(z_j - \rho\right) + \sum_{i \neq j} \exp z_i}, \tag{7.22}$$

$$\text{for } j = 1, \dots, k; \quad \rho \in \mathbb{R}_+$$

and this induces the soft-margin cross-entropy loss $\mathcal{L}_f^{(\rho)}$:

$$\begin{aligned} \mathcal{L}_f^{(\rho)}(x, y) &\triangleq -\log \sigma_y^{(\rho)}(f(x)) \\ &= \log \sum_{y' \in \mathcal{Y}} \exp\left(f_{y'}(x) - f_y(x) + \rho \cdot \mathbb{1}_{y' \neq y}\right) \end{aligned} \tag{7.23}$$

Likewise, a soft-max adversarial loss can also be defined as follows:

$$\tilde{\ell}_f^{(\rho)}(\boldsymbol{x}, y) \triangleq \log\left(1 - \sigma_y^{(\rho)}(f(\boldsymbol{x}))\right) \tag{7.24}$$

We prove this soft-margin-based bound with the help of the following lemma, which motivates us to investigate $\ell_f^{(\rho)}$ further in Section 7.4.2.4.

Lemma 7.1 *The soft-max cross entropy bounds the margin loss as follows:*

$$\Phi^{(\rho)} \circ \phi_f(\boldsymbol{x}, y) \leq \frac{1}{2\rho}\ell_f^{(2\rho)}(\boldsymbol{x}, y) \tag{7.25}$$

Proof: First, let us recall the generalized hinge loss [32, Section 17.2]:

$$\hbar_f^{(\theta)}(\boldsymbol{x}, y) \triangleq \max_{y' \in \mathcal{Y}} \left(f_{y'}(\boldsymbol{x}) - f_y(\boldsymbol{x}) + \theta \cdot \mathbb{1}_{y' \neq y}\right) \tag{7.26}$$

Noting that the argument of the max function always equals 0 for $y' = y$, we can write

$$\hbar_f^{(2\rho)}(\boldsymbol{x}, y) = \max\left\{0, \max_{y' \neq y} f_{y'}(\boldsymbol{x}) - f_y(\boldsymbol{x}) + 2\rho\right\} = \max\left\{0, 2\rho - 2\phi_f(\boldsymbol{x}, y)\right\}$$
$$= 2\rho\max\left\{0, 1 - \phi_f(\boldsymbol{x}, y)/\rho\right\} \tag{7.27}$$

The last max expression in (7.27) can be derived from the margin loss if the output of $\Phi^{(\rho)}$ for $x \leq 0$ is set to $1 - x/\rho$ instead of 1; hence,

$$\Phi^{(\rho)} \circ \phi_f(\boldsymbol{x}, y) \leq \frac{1}{2\rho}\hbar_f^{(2\rho)}(\boldsymbol{x}, y) \tag{7.28}$$

and the proof is concluded using the fact that

$$\log \sum_{a \in \mathcal{A}} \exp a \geq \max_{a \in \mathcal{A}} a \tag{7.29}$$

for any finite set $\mathcal{A}$.

Taking the expectation with regards to p_S in (7.25), we finally obtain

$$\text{err}_S^{(\rho)}(f) \leq \frac{1}{2\rho}\mathbb{E}_S\ell_f^{(2\rho)}(\boldsymbol{x}, y) \tag{7.30}$$

7.4.2.4 Setup
Similarly as in [24], we focus on human activity recognition using data that have been measured simultaneously with four different 60-GHz frequency modulated continuous wave (FMCW) radar sensors. For these measurements, two male subjects were recorded separately while performing five different activities:

Table 7.1 Radar configuration parameters.

Configuration name		I	II	III	IV
Chirps per frame	n_c	64	64	64	64
Samples per chirp	n_s	256	256	**128**	256
Bandwidth	(GHz)	2	**1**	2	2
Frame period	(ms)	**50**	32	32	32
Chirp to chirp time	(µs)	250	250	250	250
Range resolution	(cm)	7.5	**15**	7.5	7.5
Max. range	(m)	6.2	**12.5**	**4.8**	6.2
Max. speed	(m/s)	5.0	5.0	5.0	5.0
Speed resolution	(m/s)	0.15	0.15	0.15	0.15

standing, waving, walking, boxing or *boxing while walking*. Each one of the radar sensors was configured with a different set of radar parameters, presented in Table 7.1 as **I** to **IV**. Here, the divergent parameters of the different configurations (marked in bold) affect the temporal and range resolution of the RDM sequences, as well as the maximum observable scope of the latter. From these configurations, I has been taken over from [24].

The input features x in Figure 7.8 comprise both range (x_r) and Doppler (x_d) information, i.e.,:

$$x = (x_r, x_d), \quad x_r, x_d \in \mathbb{R}^{64 \times 128} \tag{7.31}$$

The radar preprocessing to produce x is also based on [24], with the notable addition of cropping and resampling of the spectrograms to ensure the dimensions in (7.31) and the scopes of 0–2 seconds 0.0–4.8 m, and −5–5 m/s for the time, range, and Doppler dimensions, respectively.

Despite all the preprocessing, the differences on resolution still yield a domain shift across configurations that we try to tackle with MDD. For that, we take both spectrograms as an input to our feature extractor ψ. Here we choose the same topology as in [24]; that is, a pair of twin branches ψ_r and ψ_d, each one consisting of three convolutional layers for which we concatenate the outputs:

$$\psi(x) \equiv (\psi_r(x_r), \psi_d(x_d)) \tag{7.32}$$

Furthermore, we employ a bottleneck layer of 512 nodes and choose our hypothesis space $\mathcal{F}$ to be consistent with the structure of the fully connected layers from [24].

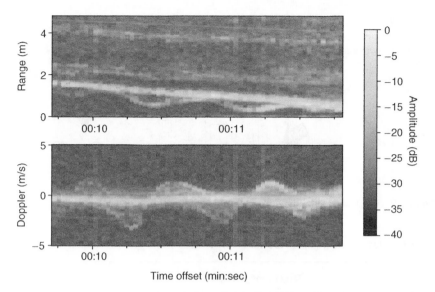

Figure 7.8 Range and Doppler spectrogram for *boxing while walking*.

Motivated by Lemma 7.1, we replace the vanilla loss terms in (7.21) by the soft-margin losses $\mathcal{L}^{(\rho)}$ and $\tilde{\mathcal{L}}^{(\rho)}$ with

$$\rho = 2\log 2 \simeq 1.386 \tag{7.33}$$

and we set the margin factor $\gamma = 1$ since the margin ρ is already included in the loss. Our version of MDD becomes thus

$$\hat{d}^{(\rho)}_{f,\psi,\mathcal{F}}\left(\hat{S},\hat{T}\right) \triangleq \max_{f' \in \mathcal{F}} \mathbb{E}_{x^t \sim \hat{T}} \tilde{\mathcal{L}}^{(\rho)}_{f'}\left(\psi\left(x^t\right), h_f\left(\psi\left(x^t\right)\right)\right)$$
$$- \mathbb{E}_{x^s \sim \hat{S}} \mathcal{L}^{(\rho)}_{f'}\left(\psi\left(x^s\right), h_f\left(\psi\left(x^s\right)\right)\right) \tag{7.34}$$

Other than that, we leave all hyperparameters to the same values as in [27] and adapt their implementation as in Figure 7.9. This has been written in Pytorch as an instance of adversarial training, where a GRL is used to minimize the MDD loss term on ψ while maximizing on f' as the minimax formulation in (7.15) mandates.

The number of samples per dataset lies over 1150 samples for the train sets and over 350 samples for the test sets.

7.4.2.5 Results

We have run unsupervised training experiments for all possible domain pairs within configurations **I–IV** and summarized the resulting test accuracies on the test sets in Table 7.2.

The figures follow the same trend as the results of MDD in computer vision datasets as reported by Zhang et al. [27] and presented in Table 7.3. Here one can

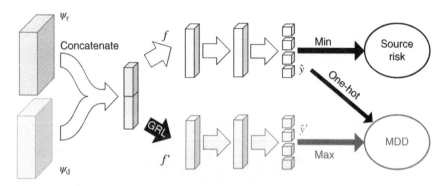

Figure 7.9 MDD adversarial network. Source: Adapted from Zhang et al. [27].

Table 7.2 Test accuracy (%) of MDD for FMCW data.

		Target configuration			
		I	II	III	IV
	I	—	91.4	90.6	88.3
Source configuration	II	90.9	—	89.8	89.4
	III	89.4	90.4	—	89.4
	IV	92.5	85.8	90.9	—

Table 7.3 Minimum and maximum accuracy (%) of MDD for different datasets.

Office-31	Office-Home	VisDa	FMCW
72.2–100.0	53.6–82.3	74.6 (single value)	85.8–92.5

compare, for instance, the results using MDD for our FMCW data with the minimum and maximum accuracies obtained for Office-31, a dataset containing 4652 images from three domains [33]. It is also noteworthy that the highest accuracy for FMCW exceeds both that of the Office-Home dataset (15 500 images from four domains) [34] and the VisDA dataset (280k real and synthetic images) [35].

Our results are also comparable with the FADA method for FMCW-based human activity recognition in [24], which increases the baseline accuracy of 50–60% without domain adaptation to 88–92%. Here it is important to note that

Table 7.4 Average accuracy comparison (%) of the original MDD implementation and the soft-margin version.

	Office-31	Office-Home	FMCW
Original MDD	88.9	68.1	89.525
Soft-margin MDD	88.3	67.6	89.9

MDD is, in contrast to FADA, an unsupervised technique, and thus, it presents the advantage of working with unlabeled target data.

We have also compared the average accuracy across domain combinations for the original implementation of MDD in (7.21) with the average accuracy for our soft-margin version in (7.34), taking the Office-31 and Office-Home datasets as well as our FMCW radar data. The results, which can be seen in Table 7.4, show little difference between both implementations.

7.4.2.6 Conclusion

In this work, we confirm that the MDD algorithm, which has already shown promising results for unsupervised domain adaptation in the area of computer vision, is also suitable for radar data across different FMCW parameters. The obtained accuracy can become as high as for some supervised techniques [24] while using a much more limited dataset, paving thus the way for a prompt deployment of radar-based deep learning applications with custom configurations.

In our experiments, we observe that the use of the soft-margin cross entropy loss provides similar results as the original implementation by Zhang et al. [27]. Since the motivation of MDD is to bring the algorithms closer to the analytical performance bounds of domain adaptation, we see potential in this alternative loss function to bridge the gap between theory and practice.

7.5 Summary

In this chapter, we have learned about the basics and more advanced methods of domain adaptation, as well as some practical examples in radar-based applications. We have seen that domain adaptation is often useful if we want to improve generalization or have only a limited amount of training data for the target domain. We have seen that there are a lot of different approaches to domain adaptation, all of which try to map the source domain and the target domain to some common space. The last part of this chapter contains some questions for the reader to check on their understanding of the content.

7.6 Questions to the Reader

- What is Domain Adaptation used for?
- What are some common data shifts?
- Why do we generally want to map samples from source domain and samples from the target domain to similar feature vectors?
- Why is there a gradient reversal step in Figure 7.5 between the encoder and the discriminator?
- For DA in radar: which other sensors are typically used for source domain data?
- What is the curse of dimensionality?
- How can MDD be used to better generalize to different FMCW radar settings?

References

1 Torrey, L. and Shavlik, J. (2010). Transfer learning. *Handbook of Research on Machine Learning Applications and Trends: Algorithms, Methods, and Techniques*. IGI Global, pp. 242–264.

2 Csurka, G. (2017). Domain adaptation for visual applications: a comprehensive survey. *arXiv preprint arXiv:1702.05374*.

3 Wang, M. and Deng, W. (2018). Deep visual domain adaptation: a survey. *Neurocomputing* 312: 135–153.

4 Yin, Y., Yang, Z., Hu, H., and Wu, X. (2021). Metric-learning-assisted domain adaptation. *Neurocomputing* 454: 268–279.

5 Kouw, W.M. and Loog, M. (2019). A review of domain adaptation without target labels. *IEEE Transactions on Pattern Analysis and Machine Intelligence* 43 (3): 766–785.

6 Wang, W., Li, H., Ding, Z., and Wang, Z. (2020). Rethink maximum mean discrepancy for domain adaptation. *arXiv preprint arXiv:2007.00689*.

7 Li, Y., Wang, N., Shi, J. et al. (2018). Adaptive batch normalization for practical domain adaptation. *Pattern Recognition* 80: 109–117.

8 Stephan, M., Stadelmayer, T., Santra, A. et al. (2021). Radar image reconstruction from raw ADC data using parametric variational autoencoder with domain adaptation. *2020 25th International Conference on Pattern Recognition (ICPR)*, pp. 9529–9536.

9 Kirkpatrick, J., Pascanu, R., Rabinowitz, N. et al. (2017). Overcoming catastrophic forgetting in neural networks. *Proceedings of the National Academy of Sciences of the United States of America* 114 (13): 3521–3526.

10 Luo, Y.-W., Ren, C.-X., and Chen, Z.-Y. (2021). Geometry-aware unsupervised domain adaptation. *arXiv preprint arXiv:2112.11041*.

11 Ganin, Y. and Lempitsky, V. (2015). Unsupervised domain adaptation by backpropagation. *International Conference on Machine Learning*. PMLR, pp. 1180–1189.

12 Riz, E., Demir, B., and Bruzzone, L. (2016). Domain adaptation based on deep denoising auto-encoders for classification of remote sensing images. *Image and Signal Processing for Remote Sensing XXII*, Volume 10004. International Society for Optics and Photonics, p. 100040K.

13 Rahman, M.M., Rahman, T., Kim, D., and Alam, M.A.U. (2021). Knowledge transfer across imaging modalities via simultaneous learning of adaptive autoencoders for high-fidelity mobile robot vision. *2021 IEEE/RSJ International Conference on Intelligent Robots and Systems (IROS)*. IEEE.

14 Peng, Z., Li, C., Mu noz-Ferreras, J.-M., and Gómez-García, R. (2017). An FMCW radar sensor for human gesture recognition in the presence of multiple targets. *2017 First IEEE MTT-S International Microwave Bio Conference (IMBIOC)*, pp. 1–3.

15 Santra, A., Ulaganathan, R.V., and Finke, T. (2018). Short-range millimetric-wave radar system for occupancy sensing application. *IEEE Sensors Letters* 2 (3): 1–4.

16 Lien, J., Gillian, N., Karagozler, M.E. et al. (2016). Soli: ubiquitous gesture sensing with millimeter wave radar. *ACM Transactions on Graphics* 35 (4): 142-1–142-19.

17 Trotta, S., Weber, D., Jungmaier, R.W. et al. (2021). 2.3 SOLI: a tiny device for a new human machine interface. *2021 IEEE International Solid- State Circuits Conference (ISSCC)*, Volume 64, pp. 42–44.

18 Santra, A. and Hazra, S. (2020). *Deep Learning Applications of Short-Range Radars*. Artech House.

19 Redko, I., Morvant, E., Habrard, A. et al. (2019). *Advances in Domain Adaptation Theory*. Elsevier.

20 Kouw, W.M. and Loog, M. (2018). An introduction to domain adaptation and transfer learning.*arXiv:1812.11806 [cs, stat]*.

21 Yin, W., Yang, X., Li, L. et al. (2019). Self-adjustable domain adaptation in personalized ECG monitoring integrated with IR-UWB radar. *Biomedical Signal Processing and Control* 47: 75–87.

22 Chen, Q., Liu, Y., Fioranelli, F. et al. (2019). Eliminate aspect angle variations for human activity recognition using unsupervised deep adaptation network. *2019 IEEE Radar Conference (RadarConf)*, pp. 1–6.

23 Li, X., Jing, X., and He, Y. (2020). Unsupervised domain adaptation for human activity recognition in radar. *2020 IEEE Radar Conference (RadarConf20)*, pp. 1–5.

24 Khodabakhshandeh, H., Visentin, T., Hernangómez, R., and Pütz, M. (2021). Domain adaptation across configurations of FMCW radar for deep learning based human activity classification. *2021 21st International Radar Symposium (IRS)*, Berlin, Germany, pp. 1–10.

25 Motiian, S., Jones, Q., Iranmanesh, S., and Doretto, G. (2017). Few-shot adversarial domain adaptation for deep classification. In: *Advances in Neural Information Processing Systems*, vol. 30 (ed. I. Guyon, U. Von Luxburg, S. Bengio, H. Wallach, R. Fergus, S. Vishwanathan, and R. Garnett), 6673–6683. Curran Associates, Inc.

26 Xu, X., Zhou, X., Venkatesan, R. et al. (2019). D-SNE: Domain adaptation using stochastic neighborhood embedding. *Proceedings of the IEEE Conference on Computer Vision and Pattern Recognition*, pp. 2497–2506.

27 Zhang, Y., Liu, T., Long, M., and Jordan, M. (2019). Bridging theory and algorithm for domain adaptation. *International Conference on Machine Learning*. PMLR, pp. 7404–7413.

28 Mansour, Y., Mohri, M., and Rostamizadeh, A. (2009). Domain adaptation: learning bounds and algorithms. *Proceedings of The 22nd Annual Conference on Learning Theory (COLT 2009)*, Montréal, Canada.

29 Mohri, M., Rostamizadeh, A., and Talwalkar, A. (2018). *Foundations of Machine Learning*, 2e. Cambridge, MA; London, England: MIT Press.

30 Goodfellow, I., Pouget-Abadie, J., Mirza, M. et al. (2014). Generative adversarial nets. In: *Advances in Neural Information Processing Systems*, vol. 27 (ed. Z. Ghahramani, M. Welling, C. Cortes et al.), 9. Curran Associates, Inc.

31 Liang, X., Wang, X., Lei, Z. et al. (2017). Soft-margin softmax for deep classification. In: *Neural Information Processing, ser. Lecture Notes in Computer Science* (ed. (ed. D. Liu, S. Xie, Y. Li), 413–421. Cham: Springer International Publishing.

32 Shalev-Shwartz, S. and Ben-David, S. (2014). *Understanding Machine Learning: From Theory to Algorithms*. Cambridge University Press.

33 Saenko, K., Kulis, B., Fritz, M. et al. (2010). Adapting visual category models to new domains. In: *Computer Vision – ECCV 2010, ser. Lecture Notes in Computer Science*, 213–226. Berlin, Heidelberg: Springer.

34 Venkateswara, H., Eusebio, J., Chakraborty, S., and Panchanathan, S. (2017). Deep hashing network for unsupervised domain adaptation. Proceedings of the IEEE Conference on Computer Vision and Pattern Recognition, pp. 5018–5027.

35 Peng, X., Usman, B., Kaushik, N. et al. (2017). VisDA: The Visual Domain Adaptation Challenge. *arXiv:1710.06924 [cs]*.

8

Bayesian Deep Learning

After reading this chapter, the reader will have an understanding on

- An overview of the principle of learning theory for both deterministic and Bayesian neural networks.
- Different elemental blocks required to formulate Bayesian deep learning and different optimization techniques.
- Application on Bayesian deep learning with its advantage for both seen and unseen data.

Deep learning has led to a revolution in machine learning by providing solution to highly nonlinear and complex problems. The concept of deep learning can be seen in two different senses. First, as a toolbox or a toolkit of methods for solving challenging problems using lots of data, i.e., by utilizing available data into one of many possible methods such as k-means, random forest, neural networks (NNs) such as convolution neural networks (CNNs), long short-term memory (LSTM), and recurrent neural network (RNN), Q-learning. The choice of method is done either from literature or by having good theoretical or empirical performance over subsampled (validation) data, whereas in another paradigm with its own limitations, deep learning is the science of learning methods or model from data. Here, model is meant as a description of all possible data one could observe from a given target system or environment but not the data which we use during training. The data used during the training are sampled information about target environment. As long as measured data are univariant and uniform, the mean value of the measured data can be approximated as data population. But in real world, measured data are multivariant and multimode. As a result, the mean value of the measured data won't give all the information about data population distribution. Thus, the learned models have limited knowledge on the environment. Additionally, often the measured data are noisy which also leads to the loss of information. Because of such situation, learned model is tied to the concept of uncertainty, i.e., to make

Methods and Techniques in Deep Learning: Advancements in mmWave Radar Solutions, First Edition.
Avik Santra, Souvik Hazra, Lorenzo Servadei, Thomas Stadelmayer, Michael Stephan, and Anand Dubey.
© 2023 The Institute of Electrical and Electronics Engineers, Inc. Published 2023 by John Wiley & Sons, Inc.

prediction about the data which model haven't seen yet, the learned model needs to take uncertainty into account. The need for model uncertainty is most critical for user–safety-based applications like self-driving cars and medical fields. This gives importance to statistical models such as Bayesian neural networks (BNNs) which help to quantify uncertainty over its estimates for a given input data which model has not been seen before.

The literature is rich about Bayesian deep learning (BDL). In order to show how BDL can help in real-life problem of short-range radar sensors, in this chapter, first we review and navigate through fundamentals of learning theory behind deterministic and statistical model.[1] Afterward, the different optimization techniques from literature are reviewed as practitioners for building BDL. Finally, a real-life use case is presented, which includes approaches of an integrated Bayesian tracker for the tracking and classification of road users, such as pedestrian and cyclist.

8.1 Learning Theory

The goal of the NN is to learn model for given measured data $D = \left\{ \left\{ \left(\mathbf{x}^{(n)}, y^{(n)} \right) \right\}_{n=1}^{N} = (X, \mathbf{y}) \right\}$, sampled from independent and identically distributed (i.i.d.).[2] Here, $\mathbf{x}^{(n)}, \mathbf{y}^{(n)}$ represents input data to the model and ground truth label of dimension n with N number of training samples. The learning is done in two stages: (i) by mapping input pattern $\mathcal{X}$ to hidden representation $\mathcal{T}(x)$, known as representation learning or encoder, and (ii) decoding hidden features to the best optimal label $\hat{y}$. This process is termed as learning theory of model and can be summarized as follows:

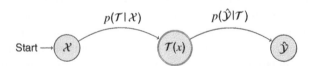

The goal of the model is to predict the best optimal output $\hat{y}$ from representation $\mathcal{T}(x)$ which is close to label (ground truth). Thus, model tries to minimize the empirical error between estimated and label output during the training. The minimization function could be a distance-based (for deterministic model) or a density-based (for statistical model) error function e, depending on model

---

1 For ease of synchronization to this chapter and narrow down our focus, NNs are considered as choice of model.
2 In probability theory and statistics, a collection of random variables is i.i.d. if each random variable has the same probability distribution as the others, and all are mutually independent.

definition. While minimizing the empirical (training) error during the training, the global goal is to minimize generalization gap too, i.e., to let model predict with bounded error for samples outside of training data, as described by Eq. (8.1).

$$\text{Generalization Gap} = \text{Empirical Error} - \text{Expected Error}$$
$$= \frac{1}{N} \sum_{i=1}^{N} (e(h, x_i) - E_x(e(h, x))) \tag{8.1}$$

The expected error (also known as generalization error) is the divergence in prediction for a given hypothesis h (i.e., $\theta_\mu, \theta_\sigma, \mathcal{M}$) when optimized over all possible data. This gets trivial to be optimized for a limited training data D.

On the other hand, for a large scale of training data, the generalization gap is minimized asymptotically. Let x_i be i.i.d from D with a probability $p(x_i)$ greater than zero, i.e., $x_i \sim p(x) > 0$. As a result, for N sample of measurement data D, the joint probability $P(x_1, \ldots, x_n)$ is equal to the product of their individual probability $\prod_{i=1}^{N} P(x_i)$. To make a product of probabilistic distribution computationally inexpensive, logarithm can be taken which makes product equivalent to summation for i.i.d, and is equal to $\sum_{i=1}^{N} \log p(x)$. As a result, with large number of measured data (asymptotically) with $N \to \infty$, the summation of probability distribution of all sampled data will concentrate around the mean of the population distribution. The mean value, in this case, is simply the expectation value $\langle -\log p(x) \rangle_x$ for all possible x which can be written as $\sum_x p(x) \lg p(x)$ and known as entropy of x. This indicates that the entropy measure of x can approximate the system for very large measured data. This further shows that the probability of finding sequences which are further away from mean (entropy of x) is equivalent to zero, as N is very large.

With the asymptotic approximation over large-scale and high-dimensional training data, the model is optimized around mean. As a result, maximum likelihood is considered as choice of error function. After optimization of model, the estimation by model over new unseen data is done using extrapolation of learned function, i.e., around mean value instead of true distribution. Although in real-world scenario, despite large-scale training data, the subclass target samples fall into smaller subspace of global input space and thus making limitations on asymptotic approximation over measured data. Additionally, this also leads to multimode optimization problem which causes large generalization error for unseen data as model fails to extrapolate due to multiple local minima. This brings the requirement to quantify uncertainty over both model and its estimation.

8.2 Bayesian Learning

To address the problem with limited training data and model, its stochastic nature in statistics provides a framework to use mathematics of probability theory to

Table 8.1 Different parametric and nonparametric models with possible algorithms to optimize them.

Parametric	Nonparametric	Algorithm	Application
Polynomial regression	Gaussian processes	MCMC	Function approximation
Logistic regression	Gaussian process classifiers	Variational Bayes	Classification
Mixture models, k-means	Dirichlet process mixtures	Stochastic gradient descent	Clustering
Hidden Markov models	Infinite HMMs	Conjugate-gradient	Time series
Factor analysis/ pPCA/PMF infinite	Latent factor models	Belief propagation	Feature discovery

express all forms of uncertainty associated with model and data. In a broad view, statistical framework can be grouped into two models – nonparametric and parametric Bayesian models. While both the approaches try to find optimum model parameter θ which can estimate true label for new input data, parametric model assumes finite set of parameters, i.e., the complexity of model is bounded even if data D is unbounded. In contrast to this, nonparametric model assumes that the data distribution cannot be defined by predefined finite set of parameters but can be learned from data by assuming an infinite dimension over θ. This is done by assuming θ as a function. Table 8.1 shows different known parametric and nonparametric models together with possible applications. It is important to note that, all the algorithm mentioned in Table 8.1 are independent of the model and are useful for model optimization/learning. In this chapter, we will have more focus on the parametric model using Bayesian algorithm as our topic of interest is deep neural networks (DNNs), whereas a small overview on nonparametric Bayesian models will be given at the end of this section.

8.2.1 Parametric Bayesian Models

In the statistical modeling of the parametric model, the stochastic nature of measured data is assumed to be sampled from a population of i.i.d. In literature, the statistical methods to learn about distribution can be grouped into two schools of thought, Bayesian (also known as inverse-probability) paradigm and frequentist paradigm. In contrast to frequentist view, the Bayesian framework considers model and its associated parameters as random (uncertain) variables which follow a probability distribution. As a result, Bayesian framework is based on two basic principles: (i) probability is a measure of belief in the occurrence of measured sampled data, rather than the limiting of it by the frequency of occurrence where the number of measured sampled data goes toward infinity for consensus, as assumed in the frequentist paradigm, and (ii) Bayesian framework allows to infer and update the hypothesis, i.e., belief of unknown quantities from measured data (evidence) using Bayes' theorem, as follows:

$$P(\text{hypothesis} \mid \text{data}) = \frac{P(\text{hypothesis})P(\text{data} \mid \text{hypothesis})}{\sum_h P(h)P(\text{data} \mid h)} \tag{8.2}$$

Fundamentally, a point estimate-based NN model can be summarized as illustrated in Figure 8.1a. The output at any layer is a nonlinear function of dot product between input and parameter weights, i.e., $y_i^{(n)} = \sigma\left(\sum_i \theta_i x_i^{(n)}\right) + \epsilon^{(n)}$. Here, ϵ is a stochastic error function with order of number of hidden units n. As stated before, both functional model $\mathcal{M}$ and its associated parameters θ (stochastic model) are considered as random variables and brought uncertainty over functional definition (choice of architecture, number of hidden layers and units, type of nonlinearity, etc.) and optimized parameters (weights, bias). Under Bayesian framework paradigm of NN, as illustrated in Figure 8.1b, both can be treated as hypothesis $p(\theta_\mu \mid \theta_\sigma, \mathcal{M})$ and thus, holds some prior belief in the form of probability distribution. These distributions are often defined using two parameters, mean θ_μ and width of distribution, i.e., variance θ_σ, if assumed as Gaussian.

The hypothesis is updated in the form of posterior $p(\theta_\mu \mid D, \theta_\sigma, \mathcal{M})$ using sampled data (D) and encodes epistemic uncertainty, i.e., uncertainty due to limited data [1]. The update of hypothesis is known as training of the network and can be summarized by Eq. (8.3) in accordance with Bayes' Theorem. The major

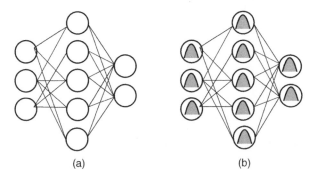

(a)   (b)

Figure 8.1 A comparative illustration of (a) deterministic and (b) Bayesian neural network.

mathematical advantage of BNN, as described by Eq. (8.3), is that it gives possibilities and probabilities in the design-structure of space $\{\theta_\mu, \theta_\sigma, \mathcal{M}\}$.

$$
\underbrace{p(\theta_\mu \mid D, \theta_\sigma, \mathcal{M})}_{\text{Posterior}} = \frac{\overbrace{p(D \mid \theta_\mu, \theta_\sigma, \mathcal{M})}^{\text{Likelihood}} \overbrace{p(\theta_\mu \mid \theta_\sigma, \mathcal{M})}^{\text{Prior}}}{\underbrace{\int p(D \mid \theta_\sigma, \mathcal{M}) p(\theta_\mu \mid \theta_\sigma, \mathcal{M}) d\theta d\mathcal{M}}_{\text{Model evidence}}} \tag{8.3}
$$

$$
\propto p(D_y \mid D_x, \theta_\mu, \theta_\sigma, \mathcal{M}) p(\theta_\mu \mid \theta_\sigma, \mathcal{M})
$$

The likelihood function ($p(D \mid \theta_\mu, \theta_\sigma, \mathcal{M})$) is a probability mass function which describes distribution over hypothesis and thus, encodes the aleatoric uncertainty, i.e., uncertainty associated with the noise in the data or model and its parameters [1]. Further, depending on the nature of noise model, aleatoric uncertainty can be categorized into heteroscedastic if noise is data-dependent and homoscedastic uncertainty where noise is data independent and uniform throughout observation. Figure 8.2 illustrates the effect of different noise observation on aleatoric uncertainty. This shows that due to inherent noise in measured data, aleatoric uncertainty cannot be reduced with more data. The likelihood function doesn't follow a probability distribution. Thus posterior, which is the multiple of likelihood and belief of the model, wouldn't follow probability

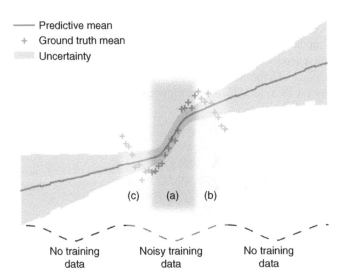

Figure 8.2 Model estimation plot on the effect of different noise observation on aleatoric uncertainty where (a) represents heteroscedastic model with data-dependent observation, (b) and (c) show homoscedastic model with small and large observation noise, respectively.

distribution. To make posterior follow nature of probability distribution, it is normalized over model evidence ($p(\mathcal{D} \mid \theta_\sigma, \mathcal{M})$).

After training, BNN uses marginal probability distribution $p\left(y \mid x, X_{\mathrm{tr}}, Y_{\mathrm{tr}}\right)$ for the prediction over new unseen data and quantifies model uncertainty on its prediction. Given $p(\theta \mid \mathcal{D})$ and $p(y \mid x, \mathcal{D})$, the prediction can be computed as follows:

$$p\left(y \mid x, X_{\mathrm{tr}}, Y_{\mathrm{tr}}\right) = \int p(y \mid x, \theta) p\left(\theta \mid X_{\mathrm{tr}}, Y_{\mathrm{tr}}\right) d\theta \tag{8.4}$$

Additionally, marginal likelihood prevents the model from over fitting on sampled training data in contrast to maximum likelihood-based optimization and inference of the model. This is due to the reason that maximum likelihood has more preference for data fitting around mean value of data distribution and may lead to over fitting. In contrast to this, marginal likelihood is an ensemble method over the possible model and its parameter which helps to estimate and to follow true distribution of data and avoid over or under fitting. For a simpler understanding, if one draws random sample from the learned posterior of the model parameter, and if model is over fitted, the maximum likelihood will give high accuracy but marginal likelihood would be very minimal. The similar approach can be used for model selection known as Bayesian razor model selection as formulated in Eq. (8.5).

$$p(\mathcal{M} \mid \mathcal{D}) = \frac{p(\mathcal{D} \mid \mathcal{M}) p(\mathcal{M})}{p(\mathcal{D})}, \quad p(\mathcal{D} \mid \mathcal{M}) = \int p(\mathcal{D} \mid \theta, \mathcal{M}) p(\theta \mid \mathcal{M}) d\theta$$
$$\tag{8.5}$$

8.2.2 Nonparametric Bayesian Models

For given training points, there could be infinitely many functions that could describe the data under bounded error gap. As a result, it is not optimal to make assumption over dimensions and number of parameters required to model the training points. With this thought, in contrast to parametric statistics, nonparametric statistics provides Bayesian framework with no assumption on the distribution of sampled data and is treated unknown by not fixing number of parameters over models. Gaussian process (GP), as nonparametric Bayesian model, provides an elegant approach toward such problem for both classification and regression. This is done by learning the probability distribution over function. The most common examples of GP are support vector machine (SVM) and kernel density estimation.

While the multivariate Gaussian captures a finite number of jointly distributed Gaussian, the GP does not have this limitation. Its mean and covariance are defined by a function. Each input to this function is a variable correlated with the other variables in the input domain, as defined by the covariance function.

Since functions can have an infinite input domain, the GP can be interpreted as an infinite dimensional Gaussian random variable (GRV). Functions from the GP require initial mean and covariance definition as prior, similar to BNNs. This shows a possible approximate relation between BNN and GP. Asymptotically, if number of hidden layers and corresponding units go to infinity with Gaussian prior, the BNN can be approximated as GP. In such situation, the nonlinearity inside BNN will be treated as kernel of GP. As a result, it is assumed that DNNs are GP. Thus, the remaining section talk about parametric Bayesian framework for DNNs.

While both Eqs. (8.3) and (8.4) show possibility to realize BNN, whereas framework has its own challenge during both training and inference stage. These challenges can be categorized into two groups: (i) initial definition of hypothesis (priors), and (ii) computational complexity to solve intractable integrals for model parameter estimation by Eq. (8.3), model estimation by Eq. (8.4), and model selection by Eq. (8.5).

8.2.3 Priors

The Bayesian framework helps to update the initial belief of the model and its parameter in the form of posterior over training data. The choice of prior controls the convergence rate to update the belief to true posterior. If the choice of prior is too strong, then update of posterior will be biased toward prior and likelihood samples will not have any effect on its update, unless the training samples are rendered independently. However, if choice of prior is very week (e.g., uniform), the posterior will predominately depend on the likelihood and one needs large training samples to reach conclusive result. The size of training samples is not well-defined in literature. As a result, a model might end up in having different posterior when initialized with different prior for limited training samples, unless they follow asymptotic consensus, i.e., optimizing Bayesian model over very large sampled training data. This shows importance on choice of prior. Based on the different set up and knowledge on experiment and data, the choice of prior can be categorized into different school of thoughts:

1. **Objective Priors:**
   Noninformative prior: Prior definition under this category does not attempt to capture any information about data. For example, if we have data which follow Gaussian distribution with mean μ and variance σ and parameter μ informs about location of the data. As a result, the choice of prior would be a distribution which doesn't capture any location information, i.e., $p(\mu) = p(\mu - a) \, \forall a$. This implies a uniform distribution which could be improper prior as it is

hard (impossible) to generalize over all parameters for a complicated model (multivariant and multimode model). Further, a similar limitation is reached when prior is picked over σ which is scaling parameter for data.

Reference prior: Prior of this kind captures the notion of noninformativeness about model parameter θ. Given a model $p(x \mid \theta)$, we wish to find the prior on θ such that an experiment involving observing x is expected to provide the most information about θ. This means most of the information about θ will come from the data rather than the prior. The information about θ can be formulated as difference between entropy of parameters θ before observing the data and entropy of parameters θ before observing the data averaged over all possible observable data:

$$I(\theta \mid x) = -\int p(\theta) \log p(\theta) d\theta - \left(-\int p(\theta, x) \log p(\theta \mid x) d\theta dx \right) \quad (8.6)$$

This can be generalized to experiments with n observations whereas prior depends on the size of data to be observed. As a result, one must know beforehand the size of data before observation which is impossible.

2. **Subjective Priors**

   These prior captures the initial belief over data and environment (e.g., tasks) which makes it very popular among Bayesian framework. To test our belief, one could generate data or task decision from the prior and can calculate expectation error. Whereas with an asymptotic assumption for a Bayesian framework, even as a vague prior belief can be useful since data will concentrate the posterior around true distribution by averaging over all possibilities. As a result, for Bayesian framework in statistical modeling, prior is not the key ingredient.

3. **Conjugate Priors**

   The conjugate priors have two advantages in a Bayesian framework: (i) it helps posterior to follow and have the same type as prior and (ii) is also useful to reduce intractable integral in Bayesian update or inference to parameter update of the prior distribution. The conjugate priors can be defined as a prior $p(\theta)$ that belongs to some distribution family A parameterized by α and likelihood function $p(D \mid \theta)$ that belongs to some distribution family B parameterized by y. Thus, if posterior $p(\theta \mid D)$ also belongs to same parametric family as prior parameterized by different parameter, then prior and likelihood are conjugate pairs.

$$p(y) \in A(\alpha), \quad p(x \mid y) \in B(y) \quad \rightarrow \quad p(y \mid x) \in A\left(\alpha'\right) \quad (8.7)$$

As a result, such priors solve the problem of intractability of full Bayesian framework in the closed form. The most common choice for conjugate pairs of distributions are the following:

Likelihood $p(x \mid y)$	y	Conjugate prior $p(y)$
Gaussian	μ	Gaussian
Gaussian	σ^{-2}	Gamma
Gaussian	(μ, σ^{-2})	Gaussian–Gamma
Multivariate Gaussian	Σ^{-1}	Wishart
Bernoulli	p	Beta
Multinomial	$(p_1, \ldots, p_m)$	Dirichlet
Poisson	λ	Gamma
Uniform	θ	Pareto

However, these conjugate prior are not suitable for multivariant and multimode model like DNNs. Additionally, to define conjugate priors, one has to know about data distribution and environment in advance which is not possible in real-world scenario.

8.3 Bayesian Approximations

In addition to choice of prior for model learning in Bayesian formulation, the key problem is the bottleneck to realize full Bayesian formulation, as defined by Eqs. (8.3) and (8.4). This is due to the computation of posterior (marginal likelihood or model evidence) and model prediction which is averaged over model parameters.

$$\text{MarginalLikelihood: } P(D \mid m) = \int P(D \mid \theta, m)P(\theta \mid m)d\theta$$
$$\text{Prediction: } P(y \mid D, m) = \int P(y \mid \theta, D, m)P(\theta \mid D, m)dm\, d\theta$$

The integrals are intractable here due to high-order interaction terms. This implies that the high-dimensional integrals required averaging over all parameters and hidden variables, making integral intractable in nature. In literature, there are multiple numerical approximation proposed to solve these integrals.

8.3.1 Laplace's Approximation

In the literature of approximation theories for intractable integrals, Laplace theory of approximation works on first-order moments, i.e., mean and variance of the true distribution. As a result, this effectively operates on an identifiable model with single mode or provides local approximation when distribution has multimode. Before looking into complex approximation for Bayesian inference, an asymptotic result for one-dimensional case will be formulated using Laplace approximation theory. For this, an intractable integral is considered in the form of

$$I(t) = \int_K h(x)e^{-tf(x)}dx \tag{8.8}$$

where K is compact (bounded and closed) subset of $\mathbb{R}^d$, and h and f are two real-valued distribution functions defined on K such that the integral is well defined for large enough $t \in \mathbb{R}$. The goal is to obtain an asymptotic equivalent of integral when t tends to infinity. In the Bayesian framework, t will be the number of observations from i.i.d. The approximation of above integral is possible iff following conditions are met:

- Continuous function: h is a continuous function on K and that f is second-order continuously differentiable function on K.
- Global minima (single mode): f has a strict global minima x_* on K, which is within the set of K, where the gradient $f'(x_*)$ is thus equal to zero, and where the Hessian[3] $f''(x_*)$ is a positive definite matrix (it is always positive semidefinite because x_* is a local minimizer of f); moreover, $h(x_*) \neq 0$.

Then as t tends to infinity, the asymptotic equivalent of integrals is as follows:

$$I(t) \sim \frac{h(x_*)}{\sqrt{\det f''(x_*)}}\left(\frac{2\pi}{t}\right)^{d/2}e^{-tf(x_*)} \tag{8.9}$$

The idea behind the above equation is quite intuitive: for $t > 0$, the exponential term $e^{-tf(x)}$ is largest when x is equal to the minima i.e., x_*. Hence, contributions that are close to x_* will only be countable in the integral. Then we can do Taylor expansions of the two functions around x_*. This results in $h(x) \approx h(x_*)$ and $f(x) \approx f(x_*) + \frac{1}{2}(x - x_*)^{\mathsf{T}}f''(x_*(x - x_*)$. Thus, approximate integral $I(t)$ as

$$I(t) \approx \int_K h(x_*)\exp\left[-tf(x_*) - \frac{t}{2}(x - x_*)^{\mathsf{T}}f''(x_*)(x - x_*)\right]dx \tag{8.10}$$

We can then make a change of variable $y = \sqrt{t}f''(x_*)^{1/2}(x - x_*)$ (where $f''(x_*)^{1/2}$ is the positive square root of $f''(x_*)$), to get, with the Jacobean[4] of the transformation leading to the term $(\det f''(x_*))^{1/2}t^{d/2}$:

$$I(t) \approx \frac{h(x_*)e^{-tf(x_*)}}{(\det f''(x_*))^{1/2}t^{d/2}}\int_{\sqrt{t}f''(x_*)^{1/2}(K-x_*)}\exp\left[-\frac{1}{2}y^{\mathsf{T}}y\right]dy \tag{8.11}$$

We can write the integral part of the expression above as follows:

$$J(t) = \int_{\mathbb{R}^d} a(y,t)dy \tag{8.12}$$

---

3 Hessian matrix is defined as second-order derivative matrix for respective function.
4 Gradient: Vector of first-order derivatives of a scalar field Jacobian: Matrix of gradients for components of a vector field Hessian: Matrix of second-order mixed partials of a scalar field.

with

$$a(y, t) = 1_{y \in \sqrt{t}K} h\left(\tfrac{1}{\sqrt{t}}y\right) \exp\left[-\tfrac{1}{2}y^\mathsf{T}y \cdot g\left(\tfrac{1}{\sqrt{t}}y\right)\right] \tag{8.13}$$

where for all $t > 0$ and $y \in \mathbb{R}^d$, $|a(y, t)| \leqslant \max_{z \in K}|h(z)| \exp\left[-\|y\|^2 \cdot \min_{z \in K}g(z)\right]$, which is integral because h is continuous on the compact set K and thus bounded, and g is strictly positive on K (since f is strictly positive except at zero as 0 is a strict global minimum), and by continuity, its minimal value is strictly positive. Thus by the dominated convergence theorem:

$$\lim_{t \to +\infty} J(t) = \int_{\mathbb{R}^d} \left(\lim_{t \to +\infty} a(y, t)\right) dy = \int_{\mathbb{R}^d} \exp\left[-\tfrac{1}{2}y^\mathsf{T}y\right] dy = (2\pi)^{d/2} \tag{8.14}$$

These results of the Gaussian integrals tend to a normalization constant of Gaussian distribution which is equal to $(2\pi)^{d/2}$. This leads to the desired result since $I(t) = J(t)/t^{d/2}$.

8.3.1.1 Laplace Approximation for Bayesian Inference

Considering the Laplace approximation for intractable integrals, as described in Eq. (8.9), both marginal likelihood during parameter update and model estimates during inference can be approximated. In this part, we have only discussed approximation for Bayesian inference, whereas a similar approach can be used for approximation of marginal likelihood. Thus, considering the integral form of Bayesian inference is as described as follows:

$$\int_\Theta h(\theta)p(\theta)\prod_{i=1}^n p\left(x_i \mid \theta\right) d\theta \tag{8.15}$$

for some function $h : \Theta \to \mathbb{R}$, are needed. For example, computing the marginal likelihood corresponds to $h = 1$. By taking logarithms over and above the formulation, we can write

$$\int_\Theta h(\theta)p(\theta)\prod_{i=1}^n p\left(x_i \mid \theta\right) d\theta = \int_\Theta h(\theta)\exp\left(\log p(\theta) + \sum_{i=1}^n \log p\left(x_i \mid \theta\right)\right) d\theta \tag{8.16}$$

and with $f_n(\theta) = -\tfrac{1}{n}\log p(\theta) - \tfrac{1}{n}\sum_{i=1}^n \log p\left(x_i \mid \theta\right)$, we have an integral in the Laplace form, that is,

$$\int_\Theta h(\theta)\exp\left(-nf_n(\theta)\right) d\theta \tag{8.17}$$

with a function f_n that now varies with n. This simple variation does not matter because of the law of large numbers, when n is large, $f_n(\theta)$ tends to a fixed function $\mathbb{E}[\log p(x \mid \theta)]$. The Laplace approximation thus requires to compute the minimizer of $f_n(\theta)$, which is exactly the maximum of a posteriori estimate $\hat{\theta}_{\text{MAP}}$, and uses the approximation:

$$\int_\Theta h(\theta) \exp\left(-nf_n(\theta)\right) d\theta \approx (2\pi/n)^{d/2} \frac{h\left(\hat{\theta}_{\text{MAP}}\right)}{\left(\det f_n''\left(\hat{\theta}_{\text{MAP}}\right)\right)^{1/2}} \exp\left(-nf_n\left(\hat{\theta}_{\text{MAP}}\right)\right)$$

(8.18)

8.3.1.2 Limitation and Extension

Although Laplace approximation helps to approximate integral of any order, it brings two major limitations. First, the approximation is done to the first order of nonlinearity which means it assumes distribution to follow single model Gaussian distribution. As a result, for a multimodel Gaussian distribution, method fails to find good approximation and leads to under-fitting. Second, as the approximation is done around the mean of the distribution considering it as a Global minima. For use case with single-mode, non-Gaussian distribution (e.g., rectangular) will result in over-fitting. Additionally, the calculation of Hessian matrix gets computationally very expensive for large model parameters. Both the limitations, high-order expansion and Hessian approximation for large model parameters are addressed in literature by following methodologies.

- The high-order expansion: The approximation is based on Taylor expansions of the functions h (order 0) and f (order 2). In order to obtain extra terms of the form $t^{d/2+v}$, for v a positive integer, we need higher-order derivatives of h and f. In more than one dimension, that quickly gets complicated (see, e.g., [2, 3]).
- The Hessian approximation: In Laplace approximation, the log of posterior is equal to summation of log of prior and log of likelihood. As number of training sample points increases, the log of likelihood increases (linearly for most model) and prior remains unaffected. As a result, the Laplace approximation can be further approximated using Bayesian information criteria (BIC) as follows:

$$\ln p(\mathbf{y} \mid m) \approx \ln p(\hat{\theta} \mid m) + \ln p(\mathbf{y} \mid \hat{\theta}, m) + \frac{d}{2} \ln 2\pi - \frac{1}{2} \ln |A| \qquad (8.19)$$

by taking the large sample limit ($n \to \infty$) where n is the number of data points:

$$\ln p(y \mid m) \approx \ln p(y \mid \hat{\theta}, m) - \frac{d}{2} \ln n \qquad (8.20)$$

8.3.2 Markov Chain Monte Carlo (MCMC)

In statistics, Markov Chain Monte Carlo (MCMC) method is used for generating samples from a given probability distribution [4]. In contrast to Laplace approximation which is limited to unimodel Gaussian distribution, sampling method can be very useful for approximation of multimodel distribution. The basic principle behind this approach is to draw samples from a probability distribution defined

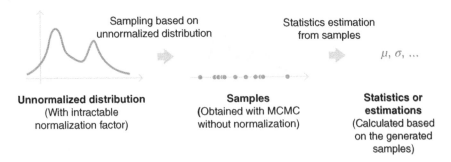

Figure 8.3 Illustration of the sampling approach.

by a factor or a rule. Later, these samples are computed to get statistical model parameters like mean and variance or approximate the distribution, as illustrated in Figure 9.20. Additionally, sampling approach also removes the assumption on modeling the true distribution using another model, in contrast to Laplace approximation and variational inference (VI) (discussed in variational approximation) (Figure 8.3). As a consequence, sampling methods result in an approximate distribution having low bias effect but at the cost of high variance. To reduce the effect of variance and more accurate approximate models, this approach is often computationally expensive.

MCMC method consists of two components: (i) the Monte Carlo part which defines sampling probability and (ii) Markov chain which defines the rule to traverse in the distribution space. Thus, MCMC builds a Markov chain whose stationary distribution is the one from where we want to sample.

- Markov Chains: It is defined as a random process over a state space E with transition probability between different substates.
- Monte Carlo methods: It is a class of algorithm that uses random sampling for computation of result. For example, an expected value of a random variable without knowing its true distribution can be evaluated empirically using a set of independent random samples.

In the context of Bayesian inference approximation, the MCMC methods are simply a class of algorithms that use Markov Chains to sample (the Monte Carlo part) from a particular probability distribution. They work by creating a Markov Chain where the limiting distribution (also known as stationary distribution) is simply the distribution we want to sample. However, not every Markov Chain has a stationary distribution or even a unique one, whereas this can be guaranteed two additional constraints to the Markov Chain:

- Irreducible: We must be able to reach any one state from any other state eventually (i.e., the expected number of steps is finite).
- Aperiodic: The system never returns to the same state with a fixed period.

Together, these two properties are known as ergodic. Another property which helps to validate if Markov Chains have stationary distribution π is known as detailed balance and reversible (also known as the detailed balance condition) Markov Chains when it satisfies below condition:

$$\pi_i P \left(X_{n+1} = j \mid X_n = i\right) = \pi_j P \left(X_{n+1} = i \mid X_n = j\right) \tag{8.21}$$

In other words, in the long run, the proportion of times that you transition from state i to state j is the same as the proportion of times you transition from state j to state i.

The Metropolis–Hastings (MH) algorithm is an MCMC technique that draws samples from a probability distribution where direct sampling is difficult. For a given target probability density function $p(x)$, the condition on MH is defined as to have a Markov chain where steady-state distribution function $f(x)$ that is proportional to $p(x)$. This gives extreme benefit in the context of Bayesian framework where marginal likelihood needs to be approximated.

Thus, to end up with steady-state distribution, Markov chain transition probabilities needs to be defined using balance condition, i.e.,

$$p(x)P \left(x \to x'\right) = p \left(x'\right) P \left(x' \to x\right) \tag{8.22}$$

Here $p(x)$ is our target distribution, and $P \left(x \to x'\right)$ is the transition probability going from point x to point x'; so our goal is to determine the form of $P \left(x \to x'\right)$. Since we get to construct the Markov Chain, let us start off by using Eq. (8.22) as the basis for that construction. As discussed before, considering the detailed balance condition guarantees that our Markov Chain has a stationary distribution and follows ergodicity (not repeating states at fixed intervals and every state much be able to reach any other state eventually), we will have built a Markov Chain that has a unique stationary distribution, $p(x)$. We can rearrange Eq. (8.22) as follows:

$$\frac{P \left(x \to x'\right)}{P \left(x' \to x\right)} = \frac{p \left(x'\right)}{p(x)} = \frac{f \left(x'\right)}{f(x)} \tag{8.23}$$

Here we use $f(x)$ to represent a function that is proportional to $p(x)$. This is to emphasize that we don't explicitly need $p(x)$, just something proportional to it such that the ratios work out to the same thing. Now, the "trick" here is that we are going to break up $P \left(x \to x'\right)$ into two independent steps: a proposal distribution $g \left(x \to x'\right)$ and an acceptance distribution $A \left(x \to x'\right)$ (similar to how rejection sampling works [2]). Since they are independent, our transition probability is just the multiplication of the two:

$$P \left(x \to x'\right) = g \left(x \to x'\right) A \left(x \to x'\right) \tag{8.24}$$

At this point, we have to figure out what an appropriate choice for $g(x)$ and $A(x)$ will be. Since $g(x)$ is the "proposal distribution," it decides the next point we will potentially be sampling. Thus, it is important that it have the same support as our target distribution $p(x)$ (ergodicity condition). A typical choice here would be the normal distribution centered on the current state. Now, given a fixed proposal distribution $g(x)$, we wish to find an $A(x)$ that matches. Rewriting Eq. (8.23) and substituting in Eq. (8.24):

$$\frac{A\left(x \to x'\right)}{A\left(x' \to x\right)} = \frac{f\left(x'\right) g\left(x' \to x\right)}{f(x)\; g\left(x \to x'\right)} \tag{8.25}$$

A typical choice for $A(x)$ that satisfies Eq. (8.25) is

$$A\left(x \to x'\right) = \min\left(1, \frac{f\left(x'\right) g\left(x' \to x\right)}{f(x)g\left(x \to x'\right)}\right) \tag{8.26}$$

We can see that by considering the cases where $\frac{f(x')g(x' \to x)}{f(x)g(x \to x')}$ is less than or equal to 1 and the cases when it is greater than 1. When it is less than or equal to 1, its inverse is greater than 1, thus, the denominator of the LHS, $A\left(x \to x'\right)$, of Eq. (8.25) is 1, while the numerator is equal to the RHS. Alternatively, when $\frac{f(x')g(x' \to x)}{f(x)g(x \to x')}$ is greater than one, the LHS numerator is 1, while the denominator is just the reciprocal of the RHS, resulting in the LHS equaling the RHS. So the overall algorithm would be the following:

(a) Initialize the initial state by picking a random x
(b) Find new x' according to $g\left(x \to x'\right)$
(c) Accept x' with uniform probability according to $A\left(x \to x'\right)$. If accepted, transition to x'; otherwise, stay in state x
(d) Go to step (b), T times
(e) Save state x as a sample, go to step (b) to sample another point

After the definition of Markov chain, the obtained sample to follow target distribution being independent needs to follow two conditions:

- Burn-in time: In order to have samples that (almost) follow the targeted distribution, we need to only consider states that are distant enough from the initialization of the generated sequence in order to reach the steady state of the Markov Chain, whereas in theory, the steady state is reached only asymptotically. Thus, the first simulated states are not usable as samples. This phase is required to reach stationary the burn-in time. Notice that, in practice, it is pretty difficult to know how long this burn-in time has to be.
- Lag sample: In order to have (almost) independent samples, we can't keep all the successive states of the sequence after the burn-in time as by definition, Markov Chain implies a strong correlation between two successive states, and thus, states that are far enough from each other to be considered as almost independent samples. In practice, the lag required between two states to be considered as almost independent can be estimated through the analysis of the auto-correlation function.

8.3.3 Variational Approximations

The major shortcoming of sampling method is the absence of objective function to measure distance between true distribution and estimation. As a result, another method called VI is proposed in the literature [3]. This method treats intractable integral approximation problem as optimization problem by defining a parameterized family of distributions and optimize it over the parameters to obtain the closest element to the target with respect to a well-defined error measure.

To ease and simplify the representation of true integrals, marginal likelihood $\ln \int p(\mathbf{y}, \mathbf{x}, \theta \mid m) d\mathbf{x} \, d\theta$ is denoted by $\pi(.)$ function. Similarly, the parameterized distribution $q(\mathbf{x}, \theta)$ is denoted by $\mathcal{F}_{\Omega}$. Thus, in more mathematical forms, the true probability distribution normalization factor C and parameterized family of distribution can be written as follows:

$$\pi(.) = C \times g(.) \propto g(.) \tag{8.27}$$

$$\mathcal{F}_{\Omega} = \{f_{\omega}; \omega \in \Omega\} \quad \Omega \equiv \text{set of possible parameters} \tag{8.28}$$

As mentioned earlier, to make approximation as optimization problem, VI uses an error measure $E()$ between two distributions and search for the best parameter such that

$$\omega^* = \underset{\omega \in \Omega}{\operatorname{argmin}} E\left(f_{\omega}, \pi\right) \tag{8.29}$$

This helps to find parameterized distribution f_{ω^*} as an approximation to true distribution π. This avoids problems such as intractable normalization integral (marginal likelihood) and combinatorics. Although VI is much simpler in comparison to MCMC, VI assumes a model in the form of parameterized family. This implies not only a bias but also a lower variance which makes VI less accurate compared to MCMC (when reached its stationary distribution), but VI produces results much faster. A visual understanding of VI is illustrated in Figure 8.4. Similar to

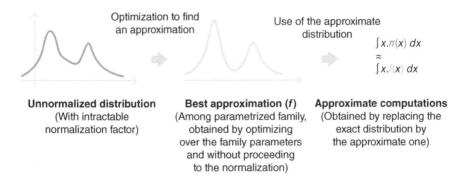

Figure 8.4 Illustration of the variational inference approximation method.

MCMC, there are three key components for VI approximation method, which are as follows: (i) family of parameterized distribution, (ii) error function, and (iii) optimization process.

- Family of parameterized distribution: The first and the most important for VI method is to set up the parameterized family of distribution which defines the search space for best approximation. The choice of the distribution family is very critical as it defines a model that controls both the bias and the complexity of the model toward approximation. If a model is very simple or have hard restriction, then it brings high bias, but the optimization process becomes very simple. In the contrast, if the model is from a complex family or have lots of free parameters, then the bias is much lower, but at the cost of complexity during its optimization. As a result, there is always a trade-off between the choice of a model to be complex enough that have less bias but also simple enough to make optimization process tractable. Additionally, if no distribution in the family is close to the target distribution, then even best optimization will result into poor approximation.

  To address this, the mean-field variational family as a choice for the family of probability distribution that is considered in [5]. The typical characteristics are that all components of the considered random vector are independent. The distributions from this family have product densities such that each independent component is governed by a distinct factor of the product. Thus, a distribution that belongs to the mean-field variational family has a density with m-dimensional random variable z that can be written as follows:

$$f(z) = \prod_{j=1}^{m} f_j\left(z_j\right) \tag{8.30}$$

Taking above equation as an example, if each density function f_j is Gaussian having mean and variance as parameters, the global density function $f(z)$ would be defined by a set of parameters from all independent Gaussian and the optimization is done over the entire set of parameters. This is visualized in Figure 8.5.

Best approximation among the family   Best approximation among the family

| Family of Gaussian distribution (F_1) with parameters: mean and variance | Dark gray – distribution for approximation, Gray – best aprroximation among F_1, Light gray – best approximation among F_2 | Family of mixture of two Gaussian distribution (F_2) with their parameters: mean and variance |

Figure 8.5 The choice of the family in variational inference sets both the difficulty of the optimization process and the quality of the final approximation.

- KL-divergence: After finding the choice of family of distribution, the way to find the best approximation of a given probability distribution raises the question as best approximation depends on the choice of error measure between approximated and true distribution. The choice of the error function should be such that the minimization problem should be sensitive to normalization factor (marginal likelihood in Bayesian inference) as the goal is to compare masses of distributions more than masses. To do so, Kullback–Leibler (KL) divergence is considered as an error function as it makes the measure insensitive to the normalization factors [6]. The KL-divergenece between two distribution (q as parameterized distribution for f_ω and p as true distribution for π) can be defined as follows:

$$D_{\mathrm{KL}}(p\|q) = \sum_{i=1}^{N} p\left(x_i\right) \cdot \left(\log p\left(x_i\right) - \log q\left(x_i\right)\right) \tag{8.31}$$

Before getting into math formulation of the KL-divergence between two distribution, it is important to understand its interpretation and properties. Essentially, Eq. (8.31) is the expectation of the log difference between the true distribution and the probability of data in the approximating distribution, i.e., Eq. (8.31) can be rewritten as $E[\log p\left(x_i\right) - \log q\left(x_i\right)]$. Despite $KL(p, q)$ matches the expectation of p better, for Bayesian approximation inverse KL-divergence ($KL(q, p)$) is estimated. This is due to the reason that KL-divergence is a moment projection (m-projection), i.e., tries to match q distribution with all the moment parameters[5] of the p distribution. To do so, one need to know p which is hard. In contrast to this, the reverse KL-divergence is information projection (i-projection) which might not yield right moments from the true distribution p but captures most information. Figure 8.6 illustrates the effect of KL-divergence and reverse KL-divergence using subplots (a–c), respectively. While the outer ring with two local optima represents true distribution, the

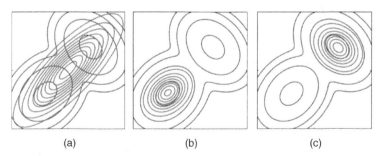

(a) (b) (c)

Figure 8.6 Visual understanding on optimization process of the variational inference approach.

***

5 Moments are model parameters which describes properties for a model distribution, e.g., mean and variance.

oval-shaped concentric circle shows approximated distribution around either side of the local optima. It is important to note that the KL-divergence is not symmetrical, thus $KL(p, q)$ is not same as $KL(q, p)$ unless q is the optimal approximate solution of p distribution.

- Optimization: Equation (8.31) can be rewritten in terms of parameterized distribution and true distribution as follows:

$$
\begin{aligned}
KL\left(f_\omega, Cg\right) &= \mathbb{E}_{z\sim f_\omega}\left[\log f_\omega(z)\right] - \mathbb{E}_{z\sim f_\omega}\left[\log(Cg(z))\right] \\
&= \mathbb{E}_{z\sim f_\omega}\left[\log f_\omega(z)\right] - \mathbb{E}_{z\sim f_\omega}\left[\log g(z)\right] - \log C \quad (8.32)
\end{aligned}
$$

This implies following the equality for error minimization problem.

$$
\omega^* = \underset{\omega\in\Omega}{\operatorname{argmin}}\ KL\left(f_\omega, \pi\right) = \underset{\omega\in\Omega}{\operatorname{argmin}}\ KL\left(f_\omega, Cg\right) = \underset{\omega\in\Omega}{\operatorname{argmin}}\ KL\left(f_\omega, g\right) \quad (8.33)
$$

Thus, having KL-divergence as a choice for an error function makes it nonsensitive to multiplicative coefficient and eases the job to estimate the best approximation from parameterized family of distribution without having to compute the normalization factor of the target distribution.

Once both the parameterized family and the error measure have been defined, we can initialize the parameters (randomly or according to a well-defined strategy) and proceed to the optimization. Several classical optimization techniques can be used, such as gradient descent or coordinate descent, that will lead, in practice, to a local optimum. In order to better understand this optimization process, let us take an example and go back to the specific case of the Bayesian inference problem where we assume a posterior such that

$$
p(z \mid x) \propto p(x \mid z)p(z) = p(x, z) \quad (8.34)
$$

In this case, if we want to get an approximation of this posterior using VI, we have to solve the following optimization process (assuming the parameterized family defined and KL divergence as error measure)

$$
\begin{aligned}
\omega^* &= \underset{\omega\in\Omega}{\operatorname{argmin}}\ KL\left(f_\omega(z), p(z \mid x)\right) \\
&= \underset{\omega\in\Omega}{\operatorname{argmin}}\ KL\left(f_\omega(z), p(x, z)\right) \\
&= \underset{\omega\in\Omega}{\operatorname{argmax}}\left(-KL\left(f_\omega(z), p(x, z)\right)\right) \\
&= \underset{\omega\in\Omega}{\operatorname{argmax}} - \mathbb{E}_{z\sim f_\omega}\left[\log \frac{f_\omega(z)}{p(z \mid x)}\right] \\
&= \underset{\omega\in\Omega}{\operatorname{argmax}}\left(\mathbb{E}_{z\sim f_\omega}[\log p(z)] + \underbrace{\mathbb{E}_{z\sim f_\omega}[\log p(x \mid z)] - \mathbb{E}_{z\sim f_\omega}\left[\log f_\omega(z)\right]}_{\text{evidence lower bound}}\right) \\
&= \underset{\omega\in\Omega}{\operatorname{argmax}}\left(\mathbb{E}_{z\sim f_\omega}[\log p(x \mid z)] - KL\left(f_\omega, p(z)\right)\right) \quad (8.35)
\end{aligned}
$$

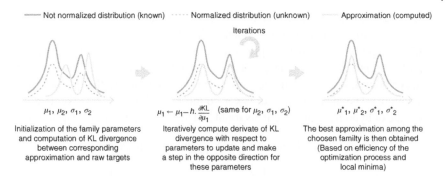

Figure 8.7 Optimization process of the variational inference approach.

The last equality helps us to better understand how the approximation is encouraged to distribute its mass. The first term is the expected log–likelihood that tends to adjust parameters so that to place the mass of the approximation on values of the latent variables z that explain the best the observed data. The second term is the negative KL divergence between the approximation and the prior that tends to adjust the parameters in order to make the approximation be close to the prior distribution. Thus, this objective function expresses pretty well the usual prior/likelihood balance (Figure 8.7).

Variational Bayesian learning: Using the variational approximation (as described above), Bayesian inference for latent variable x, observed data y, and parameter θ can be approximated using lower bound on marginal likelihood as follows:

$$
\begin{aligned}
\ln p(\mathbf{y} \mid m) &= \ln \int p(\mathbf{y}, \mathbf{x}, \theta \mid m) d\mathbf{x}\, d\theta \\
&= \ln \int q(\mathbf{x}, \theta) \frac{p(\mathbf{y}, \mathbf{x}, \theta \mid m)}{q(\mathbf{x}, \theta)} d\mathbf{x}\, d\theta \\
&\geq \int q(\mathbf{x}, \theta) \ln \frac{p(\mathbf{y}, \mathbf{x}, \theta \mid m)}{q(\mathbf{x}, \theta)} d\mathbf{x}\, d\theta.
\end{aligned}
\tag{8.36}
$$

Here, $q(\mathbf{x}, \theta)$ is the parameterized approximated distribution. The inequality sign holds true using Jensen's inequality, which states, for a convex function f:

$$
f(\ln[X]) \leq \ln[f(X)]
\tag{8.37}
$$

i.e., log of weighted avg of any two points is greater than or equal to weighted avg of log of those two points. This holds in any number of dimensions. This helps to find lower bound on the evidence (marginal likelihood). Further, using factorized approximation on parameterized distribution $q(\mathbf{x}, \theta) \approx q_{\mathbf{x}}(\mathbf{x}) q_{\theta}(\theta)$, Eq. (8.36) can be rewritten as follows:

$$
\begin{aligned}
\ln p(\mathbf{y} \mid m) &\geq \int q_{\mathbf{x}}(\mathbf{x}) q_{\theta}(\theta) \ln \frac{p(\mathbf{y}, \mathbf{x}, \theta \mid m)}{q_{\mathbf{x}}(\mathbf{x}) q_{\theta}(\theta)} d\mathbf{x}\, d\theta \\
&\stackrel{\text{def}}{=} \mathcal{F}_m \left(q_{\mathbf{x}}(\mathbf{x}), q_{\theta}(\theta), \mathbf{y} \right)
\end{aligned}
\tag{8.38}
$$

As a result, maximizing lower bound will make our parameterized distribution close to true distributions. Maximizing $\mathcal{F}_m$ is equivalent to minimizing KL-divergence between the approximate posterior, $q_\theta(\theta)q_x(\mathbf{x})$ and the exact posterior, $p(\theta, \mathbf{x} \mid \mathbf{y}, m)$, whereas it is important to note that without having upper bound, the approximation may lead to over fitting (although very rare situation). The maximization of lower-bound $\mathcal{F}_m$ is done in *EM*-alike iterative update of hypothesis and hidden variables. This can be summarized as follows:

$$q_{\mathbf{x}}^{(t+1)}(\mathbf{x}) \propto \exp\left[\int \ln p(\mathbf{x}, \mathbf{y} \mid \theta, m) q_\theta^{(t)}(\theta) d\theta\right] \quad \text{E-like step}$$

$$q_\theta^{(t+1)}(\theta) \propto p(\theta \mid m) \exp\left[\int \ln p(\mathbf{x}, \mathbf{y} \mid \theta, m) q_{\mathbf{x}}^{(t+1)}(\mathbf{x}) d\mathbf{x}\right] \quad \text{M-like step}$$

$$(8.39)$$

The E-like step updates the distribution at time $t + 1$ over latent variables x which is a function of distribution over parameter θ at time t. At next step, M-like step, the distribution of parameter θ at time $t + 1$ is updated which is a function of distribution of latent variables at time t.

EM for MAP estimation Goal: maximize $p(\theta \mid \mathbf{y}, m)$ w.r.t. θ E step: compute

$$q_{\mathbf{x}}^{(t+1)}(\mathbf{x}) = p\left(\mathbf{x} \mid \mathbf{y}, \theta^{(t)}\right) \tag{8.40}$$

M step:

$$\theta^{(t+1)} = \underset{\theta}{\text{argmax}} \int q_{\mathbf{x}}^{(t+1)}(\mathbf{x}) \ln p(\mathbf{x}, \mathbf{y}, \theta) d\mathbf{x} \tag{8.41}$$

Variational Bayesian EM Goal: lower-bound $p(\mathbf{y} \mid m)$ VB-E step: compute

$$q_{\mathbf{x}}^{(t+1)}(\mathbf{x}) = p\left(\mathbf{x} \mid \mathbf{y}, \overline{\phi}^{(t)}\right) \tag{8.42}$$

VB-M step:

$$q_\theta^{(t+1)}(\theta) \propto \exp\left[\int q_{\mathbf{x}}^{(t+1)}(\mathbf{x}) \ln p(\mathbf{x}, \mathbf{y}, \theta) d\mathbf{x}\right] \tag{8.43}$$

8.4 Application: VRU Classification

In this section, we will demonstrate BDL with an application for Automotive radar. This will help us in understanding on approximate formulation of Bayesian framework VI, discussed in "variational Bayesian learning". The application shows the advantages of Bayesian integrated classification and tracking of vulnerable road users (VRUs) like pedestrian and cyclist. As the main focus of this chapter is on BDL, section will discuss how variational auto-encoder (VAE) and tracker follows Bayesian inference framework. At the end, we show the integration of target feature and associated noise in the form of uncertainty inside a tracker to help in association and better classification. The details on radar-specific signal processing can be found in [7, 8].

8.4.1 VAE as Bayesian

Before talking about VAE, an auto-encoder (AE) is defined as NN in the setting of encoder and decoder to learn the best encoding–decoding scheme using an iterative optimization process. In contrast to this, the VAE's principle can be defined as an auto-encoder which is trained to ensure regularity property in latent space in such a way that it enables generative process. This is done by mapping the encoder to a distribution over latent space instead of a single point. A high-level architectural difference between AE and VAE is illustrated in Figure 8.8

For recap, we denote our data with variable x which is generated from a latent/hidden variable z (the encoded representation) that is not directly observed. In VAE, contrarily to deterministic AE, both encoder and decoder are defined in probabilistic ways, $p(x \mid z)$ and $p(z \mid x)$, respectively. While the encoder describes the distribution of the hidden encoded (latent) variable given input data (also termed as posterior), decoder describes the distribution of decoded variable given hidden variables (also referred as likelihood). Following Bayes theorem and as discussed before, posterior is linked to prior $p(z)$ and likelihood.

$$p(z \mid x) = \frac{p(x \mid z)p(z)}{p(x)} = \frac{p(x \mid z)p(z)}{\int p(x \mid u)p(u)du} \tag{8.44}$$

The VAE with only deterministic loss function will ignore the encoded distribution. This makes the encoder to return either distribution with tiny variance

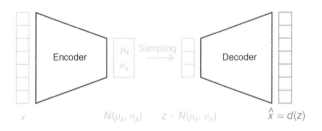

$$\text{Loss} = \| x - \hat{x} \|^2 + \text{KL}[\, N(\mu_x, \sigma_x), N(0, I)\,] = \| x - d(z) \|^2 + \text{KL}[\, N(\mu_x, \sigma_x), N(0, I)\,]$$

Figure 8.8 Illustration of difference between AE (deterministic) and VAE (probabilistic) with its loss function.

(similar to deterministic AE) or return distribution with varying mean which would break the continuity in the latent space causing poor generative behavior. In order to avoid these effects, regularization of both covariance matrix and the mean of the distributions returned by the encoder is needed. In practice, this regularization is done by enforcing distributions to be close to a standard normal distribution (centered and reduced). This way we require the covariance matrices to be close to the identity, preventing punctual distributions, and the mean to be close to 0, preventing encoded distributions to be too far apart from each other.

With this, we make an assumption on latent prior $p(z)$, posterior $p(z \mid x)$, and likelihood $p(x \mid z)$. While $p(z)$ follows standard Gaussian distribution, $p(x \mid z)$ is assumed to be Gaussian distribution with mean defined by function of latent variable $f(z)$, and covariance matrix with positive multiplicative variable c to identity matrix I. The function f is assumed to belong from family of function F.

$$p(z) \equiv \mathcal{N}(0, I)$$
$$p(x \mid z) \equiv \mathcal{N}(f(z), cI) \quad f \in F \quad c > 0 \tag{8.45}$$

As we have assumed, both $p(z)$ and $p(x \mid z)$ follow Gaussian distribution, and if $E(x \mid z) = f(z) = z$, this would imply that $p(z \mid x)$ should also follow a Gaussian distribution. As a result, in theory, one could also express $p(z \mid x)$ only in terms of the mean and the covariance matrix with respect to the means and the covariance matrices of $p(z)$ and $p(x \mid z)$. However, in practice, this condition is not met and thus, we need an approximation technique. In this chapter, we have used VI that makes the approximation more robust to some changes in the hypothesis of the model. As a result, to avoid intractable integral in Eq. (8.44), we approximate $p(z \mid x)$ by parameterized Gaussian distribution $q_x(z)$. The mean and covariance of parameterized distribution is defined by two functions of parameter x, $g(x)$, and $h(x)$. Here, both g and h belongs to the family of parameterized function.

$$q_x(z) \equiv \mathcal{N}(g(x), h(x)) \quad g \in G \quad h \in H \tag{8.46}$$

Using the concept of VI approximation, the optimization is done by minimizing KL-divergence between true $p(z \mid x)$ and parameterized distribution $q_x(z)$ to find optimal parameters (i.e., mean g^* and covariance h^*), as described below:

$$(g^*, h^*) = \underset{(g,h) \in G \times H}{\text{argmin}} \ \mathrm{KL}\left(q_x(z), p(z \mid x)\right)$$
$$= \underset{(g,h) \in G \times H}{\text{argmax}} \left(\mathbb{E}_{z \sim q_x}(\log p(x \mid z)) - \mathrm{KL}\left(q_x(z), p(z)\right) \right) \tag{8.47}$$

The second last line in the above equation, is also termed as ELBO maximization, shows two trade-off when approximation posterior $p(z \mid x)$ – either maximizing likelihood of the observation or being close to prior distribution.

This Bayesian trade-off is solved asymptotically when number of input data points is very large (infinity), whereas in a general situation, if approximation is ruled by prior, then the performance of generative decoder depends highly on the choice of function f which governs the decoder distribution. In other words, the decoder function needs to optimized in such a way that the generated output is close enough to real input when the input to function is z which is sampled from hidden parameterized distribution $q_{\tilde{x}}^*$.

$$
\begin{aligned}
f^* &= \underset{f \in F}{\mathrm{argmax}} \mathbb{E}_{z \sim q_{\tilde{x}}^*}(\log p(x \mid z)) \\
&= \underset{f \in F}{\mathrm{argmax}} \mathbb{E}_{z \sim q_x^*} \left(-\frac{\|x - f(z)\|^2}{2c} \right)
\end{aligned}
\tag{8.48}
$$

Putting all equation together, optimization of VAE can be rewritten as follows:

$$
(f^*, g^*, h^*) = \underset{(f,g,h) \in F \times G \times H}{\mathrm{argmax}} \left(\mathbb{E}_{z \sim q_x} \left(-\frac{\|x - f(z)\|^2}{2c} \right) - \mathrm{KL}\left(q_x(z), p(z) \right) \right)
\tag{8.49}
$$

where the first term measures reconstruction error and the second gives regularization on latent. One can also notice the covariance matrix term c which rules the balance between reconstruction and regularization. If the variance is high around probabilistic decoder, optimization is more favored toward regularization term.

8.4.1.1 Re-parameterization Trick

In practice (VAE), both the mean g and the covariance h functions are not completely but are also a function of common encoder, as illustrated in Figure 8.9. By definition, the covariance matrix h of approximated distribution $q_x()$ is supposed to be square, whereas for ease of computation by reducing the number of parameters, the $q_x()$ is assumed to be multidimensional Gaussian distribution with a diagonal covariance matrix, i.e., all variables are mutually independent. Thus, both $h()$ and $g()$ are vector elements of the same size. This further helps to sample latent variables for a decoder and generate new samples.

The sampling process from the latent distribution is done randomly as the learned latent embedding is a random variable which follows Gaussian distribution with mean g and covariance h. However, random sampling created a bottle for error back-propagation during optimization of the parameters. For this purpose, a method called reparameterization trick is used to enable gradient descent possible despite random sampling.

$$
z = h(x)\zeta + g(x) \quad \zeta \sim \mathcal{N}(0, I)
\tag{8.50}
$$

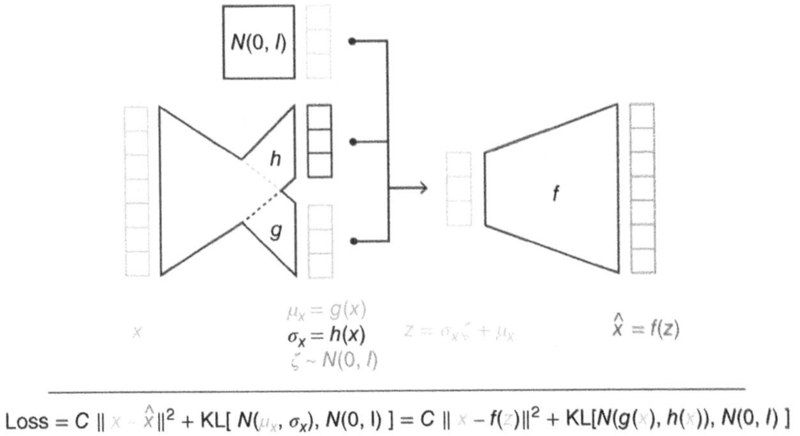

$$\mu_x = g(x)$$
$$\sigma_x = h(x)$$
$$\zeta \sim N(0, I)$$
$$z = \sigma_x \zeta + \mu_x$$
$$\hat{x} = f(z)$$

Loss = $C \parallel x - \hat{x} \parallel^2 + \text{KL}[\, N(\mu_x, \sigma_x), N(0, I)\,] = C \parallel x - f(z) \parallel^2 + \text{KL}[N(g(x), h(x)), N(0, I)\,]$

Figure 8.9 Illustration of difference between AE (deterministic) and VAE (probabilistic) with its loss function.

8.4.2 Bayesian Metric Learning

The regularization term in VAE enables Bayesian inference over latent embedding z of the network for a given input x and makes network generative in nature by preventing the model to encode data far apart in the latent space. This is done by creating a gradient over the information encoded in the latent space resulting into continuity between features. This approach is very good for representation learning but not for similarity function learning. As a result, it is very challenging to uniquely identify inter- and intraclass targets among VRU in the context of radar [7].

8.4.2.1 Vulnerable Road Users (VRUs)

The full pedestrian model is shown in Figure 8.10, where joints are depicted as dark black circles, whereas the scattering points are illustrated as dark gray crosses. To the right of the pedestrian, the cyclist model is displayed, where points with linear velocity are marked by light gray diamonds and dynamic points with rotational velocities are marked by gray circles.

Examples of relative velocities obtained from the described motion models are shown in Figure 8.10, both road users move with constant velocities along a linear trajectory toward the radar sensor. The velocities in Figure 8.10a are obtained from a male (solid lines) and a female (dashed lines) pedestrian model. The pedestrians are walking with 1.0 m/s toward the radar sensor, noticeable by the oscillating pattern around this velocity value. Differences from the gender-based gait are visible as distortions as well as different maximum velocities. However, the absolute

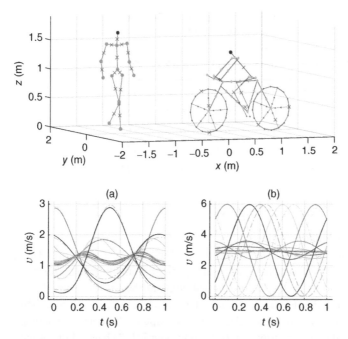

Figure 8.10 Dynamic VRU point target models for a pedestrian and a cyclist. The pedestrian joint positions are displayed as dark black circles, while static points of the cyclist model are shown by light gray diamonds and dynamic points are highlighted as dark black dots. All scattering positions are marked as dark gray crosses. Relative velocities of the VRU scattering points for a female (dashed) and a male (solid) pedestrian (a) walking with [7].

differences are relatively small and only in the order of 0.1 m/s. Apart from that Figure 8.10b shows the different radial velocities of the cyclists' scattering points, with a constant velocity of $v = 3.0$ m/s of the frame and body, as well as faster radial velocities for the wheel spokes, pedals, and legs.

For better understanding over similarity on target's appearance model, the structural similarity index measure (SSIM) over all permutations of sample classes are calculated. SSIM values closer to 1 indicate a high similarity between two images, and is in contrast to, e.g., MSE more robust to noisy variations of the image. Results on our dataset show a strong visual similarity among intraclass samples as well as between interclass targets, as indicated in Table 8.2. Most similarity indices lie around 0.5–0.7, without noticeable differences between cyclist and pedestrian class combinations. In general, this demonstrates the complexity of the problem and the importance of finding the optimum feature embeddings, which can be used for distinct appearance modeling of targets.

Table 8.2 Similarity indices (SSIM) for simulated micro-Doppler spectra of inter- and intraclasses for different VRU targets [7].

–	mPed	fPed	nPed	Cyc1	Cyc2	Cyc3
mPed	1.0	0.71	0.68	0.57	0.60	0.61
fPed	—	1.0	0.70	0.64	0.62	0.66
nPed	—	—	1.0	0.61	0.61	0.63
Cyc1	—	—	—	1.0	0.52	0.55
Cyc2	—	—	—	—	1.0	0.53
Cyc3	—	—	—	—	—	1.0

8.4.2.2 Metric Learning

Due to the strong structured similarity, we combine the idea of VI over latent embedding in combination with metric and representational learning. This helps the network to learn both the feature embedding and the variance over it. As a result, the learnt input noise can be integrated inside a tracker, which further helps during the data association (discussed later in this chapter). The details on network architecture and data preparation can be found under [7]. To have a fair comparison and similarity between the different metric learning-based VAE, Figure 8.11 is illustrated to summarize different architectures.

- Triplet variational autoencoder (TVAE): The architecture is optimized using a triplet loss [9, 10]. Prior to the network training using the triplet loss, triplet pairs are selected. These pairs consist of anchor sample (x_a), i.e., any random sample, positive sample (x_p), which is from the same class as the anchor, and a negative samples (x_n), which is a sample from any different class in comparison to the anchor class. The loss function is computed over feature embedding, i.e., the latent space (z), as shown in Eq. (8.51).

$$\mathcal{L}_{\text{triplet}} = \max \left(\|q_\phi(x_a) - q_\phi(x_p)\|^2 - \|q_\phi(x_a) - q_\phi(x_n)\|^2 \right.$$
$$\left. + \alpha_{\text{margin}}, 0 \right) \tag{8.51}$$

The network is trained following a min–max distance learning between the triplet pairs. While the distance between the anchor and negative samples is maximized by making $d(q_\phi(x_a), q_\phi(x_p)) + \alpha_{\text{margin}}$ less than $d(q_\phi(x_a), q_\phi(x_n))$, the distance between the anchor and positive samples is minimized forcing $d(q_\phi(x_a), q_\phi(x_p))$ to 0. Here, α_{margin} is a hyperparameter which defines the boundary condition between the similar and dissimilar pairs. For this process, a Euclidean distance function is considered from the available similarity metric function.

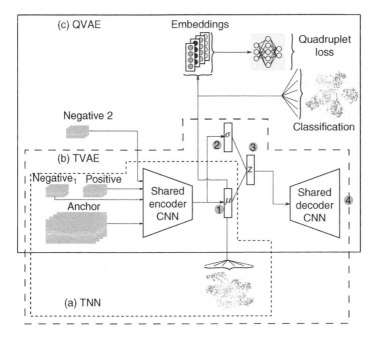

Figure 8.11 An overview of evaluated architectures and their relation to each other. While all architecture follows Bayesian fully connected network (BFCN) classifier over mean embedding for target classification, the variational inference brings additional Bayesian knowledge over extracted mean using (b) TVAE architecture in contrast to (a) TNN. With introduction of additional negative sample from similar class-group, (c) QVAE improves distinct feature learning. This is done by learning the distance between feature embedding using MLP in contrast to usage of normal L2-Norm [7].

The choice of the input triplet pair plays an important role in learning the feature embedding. Considering the spatial complexity and similarity between training examples, the mining of input triplet pairs is done considering hard examples, i.e., triplets, where the negative sample is closer to the anchor than to the positive $(d(q_\phi(x_a), q_\phi(x_n)) < d(q_\phi(x_a), q_\phi(x_p)))$ and semihard triplets, where $d(q_\phi(x_a), q_\phi(x_p)) < d(q_\phi(x_a), q_\phi(x_n)) < d(q_\phi(x_a), q_\phi(x_p)) + \alpha_{margin}$. The triplet mining is done in an online approach, i.e., during the network training.

As a result, Eq. (8.52) gives a mathematical overview on the total loss which is a linear combination of the CE loss, KL-divergence, and the triplet loss for TVAE architecture.

$$\mathcal{L}_{TVAE} = 0.7 * \mathcal{L}_{reconstruction} + 0.3 * (\mathcal{L}_{KL} + \mathcal{L}_{triplet}) \tag{8.52}$$

- Quadruplet-variational autoencoder (QVAE): TVAE suffers from two major drawbacks. First, the distance metric function for an anchor is optimized with respect to the positive and negative samples. As a result, there is no discriminator part in the triplet loss function which can help to push target samples from an intraclass. This problem is avoided by including another negative sample, belonging to the same group as the first negative sample. This helps the network to have a better inter- and intraclass distance by adding an extra parameter optimization to separate the negative class from each other. The resulting new loss function is termed as quadruplet loss ($\mathcal{L}_{\text{quadruplet}}$) and can be summarized by Eq. (8.53). It includes another hyperparameter α_2 which is kept to 0.5 during the training. While sample s_i and s_j belong to the same class and represent an anchor and positive sample, s_k and s_l belong to two different classes, which are also not an anchor class.

$$\mathcal{L}_{\text{quadruplet}} = \sum_{i,j,k}^{N} \left[q(x_i, x_j)^2 - q(x_i, x_k)^2 + \alpha_1 \right]$$

$$+ \sum_{i,j,k,l}^{N} \left[q(x_i, x_j)^2 - q(x_l, x_k)^2 + \alpha_2 \right]$$

$$s_i = s_j, s_l \neq s_k, s_i \neq s_l, s_i \neq s_k \tag{8.53}$$

The distance metric is computed with a $L2$ norm (Euclidean) which compares the feature embedding vectors elementwise with an uniform weighting to each values. Considering the nature of the training data, i.e., micro-Doppler–based signatures from VRUs, small changes in Doppler frequency from the intraclass lead to a unique identification of the target. At the same time, uniform weighting fails to find outliers within a small difference of the feature embedding which is usually the case for intraclass VRUs. Thus, the distance function is learned during the network training. For this purpose, a three-layered multilayer perceptron (MLP)-based architecture is designed and optimized using the principle of Siamese networks.

8.4.3 Kalman as Bayesian

For most general real-world (nonlinear, non-Gaussian) systems, the multidimensional Bayesian recursion becomes intractable and therefore, approximations have to be used. This includes methods such as Gaussian approximations (extended Kalman filters [EKFs]), hybrid Gaussian methods (score function EKF, Gaussian sum filters), direct and adaptive numerical integration (grid-based filters), sequential Monte Carlo methods (particle filters), and variational methods (Bayesian mixture of factor analyzers). Especially, the EKF, a suboptimal approximation of the recursive Bayesian framework, applied to a GRV of a nonlinear state, is widely used. It approximates and propagates the state distribution through the first-order Taylor series linearization, which expands the nonlinear state around a single-point. As a result, the EKF is not able to capture the uncertainty of the distribution, introducing large errors in the estimation of the true posterior mean and covariance, respectively. Alternatives can be unscented Kalman filters

(UKFs), which use deterministic sampling filters, i.e., a sigma-point Kalman filter (SPKF), to approximate the GRV by a minimal set of sample points. These sample points can capture the true mean and covariance of the GRV.

The state-of-the-art tracking algorithms for automotive use-cases mainly focus on single modalities by having target's localization information as a state vector [11, 12]. Additional target parameters such as Doppler spectra are either ignored or computed separately. In this chapter, we use both the localization and appearance model for the tracking of detected targets, by augmenting the state-vector target features, as illustrated in Figure 8.12.

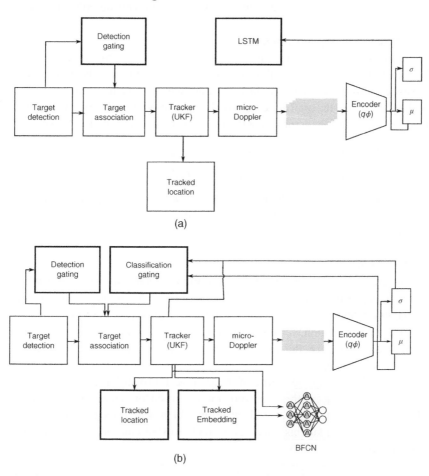

Figure 8.12 Comparison of (a) state-of-the-art algorithm pipeline for continuous tracking and classification in contrast to (b) proposed BayesRadar where additional Bayesian knowledge on extracted embedding enables the framework to follow the complete Bayesian inference [8].

As a result, the robustness of the framework is further improved by leveraging the Bayesian information associated with the input and predicted state vector and by performing data association. The framework includes multiple processing blocks of which target detection block provides measurement data on the target's localization ($Z^m(\mu, \sigma)$) to the tracker. The encoder block ($q(\phi)$) extracts appearance embedding ($E^m(\mu, \sigma)$) and augments the tracker state vector with it for each frame. The tracker (an UKF in our case) uses this information to estimate the new position of the target and classifies the target into the defined category using a Bayesian fully connected network (BFCN) classifier. The integration of the appearance model together with the gating and data association are described below.

8.4.3.1 Integrated Bayesian Tracking

Thus, the integration of the classifier output into the tracker facilitates the processing in obtaining not only the value of the current state of the classification but also the uncertainty associated with the state. Considering μ_i as the mean embedding of class i with M as the total dimension of the assumed embedding vector, the modified augmented state vector (x_a) of the tracker can be represented as Eq. (8.54).

$$x_a = \begin{bmatrix} P_x & P_y & v & Az & \mu_{11} & \mu_{12} & \cdots & \mu_{1M} \end{bmatrix}^T$$
$$g(x_a) = \begin{bmatrix} p_x^P & p_y^P & v^P & Az^P & \mu_{11}^P & \cdots & \mu_{1M}^P \end{bmatrix}^T \tag{8.54}$$

Target's localization parameter is represented by lateral (P_x), longitudinal position (P_y), velocity (v), and azimuth angle (Az). Even though Az can be estimated from P_x and Py, Az is chosen to be part of state vector.

8.4.3.2 Target Association

The accuracy of data association relies on the choice of the distance metric which can be grouped into Bayesian or non-Bayesian based on the nature of the data. Both measurement ($y_{k|k-1}^{(i)}$) and sigma-point transformed prediction ($x_{k|k-1}^{(i)}$) follow a Gaussian distribution, having a mean ($\hat{y}, \hat{x}$) and a covariance ($P_{k|k-1}^y, P_{k|k-1}^x$), respectively. Therefore, the variance over posterior and observation is used for the data association.

Additionally, due to the nature of the state vector (distribution than point), a Mahalanobis distance as the association metric is used for the computing distance. This acts as a multivariate Euclidean norm which is described in Eq. (8.55). It shows that the Mahalanobis distance is a function of both the mean and covariance of the predicted state vector.

$$d = \sqrt{(\hat{x}_{k|k-1}^{(i)} - y_{k|k-1}^{(i)})^T P_{k|k-1}^{y_i}{}^{-1}(\hat{x}_{k|k-1}^{(i)} - y_{k|k-1}^{(i)})} \tag{8.55}$$

Here, $\hat{x}_{k|k-1}^{(i)}$ is the current measurement and $y_{k|k-1}$, $P_{k|k-1}^{y_i}$ are the mean and process covariance model of the predicted state vector at a particular time step. The distance d is chi-square distributed with n_z degrees of freedom, where n_z is the dimension of the state vector which is 4 for localization and 16 for feature embedding. The measurement is associated with a particular track state only if the Mahalanobis distance is lesser than a chosen threshold. The new augmented state brings two different modalities (motion and appearance) into consideration. Thus, different thresholds for each modality are modeled which in return improves the gating operation. Overall, a threshold of 0.75 for the localization and 2.5 for the appearance model is considered. This helped to remove noisy outliers for target's localization and feature embedding (used for classification) from being associated with the states of the tracker.

8.4.4 Results

This section investigates on the accuracy and reliability for LSTM and BayesRadar toward pretrained and unseen target classes. This is done using both qualitative and quantitative analysis.

8.4.4.1 Pretrained Target Class

Classification Figure 8.13 compares the classification accuracy of VRUs (6-class) from state-of-the-art and BayesRadar using t-distributed stochastic neighbor embedding (t-SNE). To ensure better generalization of BayesRadar, results are evaluated using feature extractor network optimized with TVAE and QVAE. In contrast to Figure 8.13a–d shows better feature clustering of target class and demonstrates the advantage of QVAE over TVAE-based training. Further, the integrated framework with LSTM compared to BayesRadar shows a comparable classification accuracy of 99.45% and 99.99%, respectively, for QVAE and 93.6% and 93.3%, respectively, for TVAE.

Additionally, a quantitative analysis over t-SNE clusters, from Figure 8.13, is evaluated using silhouette and Davies–Bouldin scores. While silhouette coefficient gives a similarity measure between a sample and its own cluster (cohesion) in comparison to other clusters (separation), Davies–Bouldin coefficient indicates the distance between clusters by estimating the distance of a sample between

(a) TVAE-LSTM   (b) TVAE-BayesRadar   (c) QVAE-LSTM   (d) QVAE-BayesRadar

Figure 8.13 A visual illustration of target classification accuracy using t-SNE plot over the feature embedding tracked BayesRadar and LSTM. While QVAE-based feature extractor shows better performance over TVAE, BayesTracker shows improved class separability in tracking feature embedding in comparison to LSTM [8].

Table 8.3 Quantitative analysis on the quality of the clustering over pretrained classes estimated by encoder-LSTM and BayesRadar [8].

Clustering metric	TVAE-LSTM	TVAE-BayesRadar	QVAE-LSTM	QVAE-BayesRadar
Silhouette score	0.37	**0.41**	0.61	**0.71**
Davies–Bouldin score	1.03	**0.97**	0.45	**0.40**

The bold value here reflects the result from our proposed framework in comparison to the SOTA framework.

with-in and the neighboring clusters. Similar to the t-SNE plot illustrated in Figure 8.13, Table 8.3 shows better clustering scores for BayesRadar in comparison to LSTM. Additionally, QVAE shows better results over TVAE for both LSTM and BayesRadar.

Class Uncertainty The reliability for the demonstrated classification accuracy is evaluated by BFCN using the estimated covariance over the predicted latent embedding. As a result, uncertainty over predicted target class for BayesRadar is evaluated using the variance estimated by the UKF. In contrast to the BayesRadar, uncertainty over predicted target class by LSTM is estimated using predicted variance from pretrained variational encoder network. Figure 8.14a illustrates the

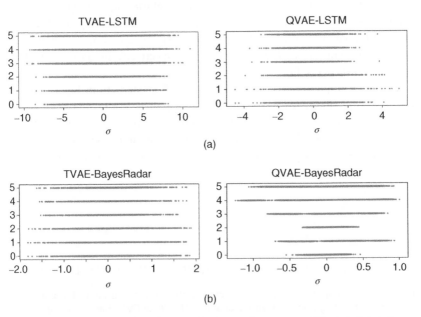

(a)

(b)

Figure 8.14 (a) Visualization of the uncertainty estimates over different classes from Encoder-LSTM (TVAE and QVAE), and (b) proposed BayesRadar [8].

learned variance, σ, (x-axis) over the target class (y-axis) using the latent embedding extracted from TVAE- and QVAE-based encoder. Figure 8.14b, evaluated on estimated latent-model from BayesRadar, shows a reduction in the variance for classification of pretrained target class. This further confirms the advantage of the proposed BayesRadar over LSTM by illustrating reduction in class uncertainty.

8.4.4.2 Unseen Target Class

Similar to Figure 8.13, t-SNE plot is used for the visualization of the new unseen class clusters. The unseen class, labeled as 6, is created by linear combination of classes 1, 3, and 4.

Classification Figure 8.15 shows the feature embedding over the new class (shaded gray) for BayesRadar and LSTM framework, having no prior learned latent model for the unseen class. Both QVAE-LSTM and QVAE-BayesRadar show better distinction against the unseen class in comparison to TVAE-LSTM and TVAE-BayesRadar framework. This shows the advantage of quadruplet-based metric learning over triplet learning, as described in [7]. The classification accuracy for QVAE-LSTM is reduced by 2.5% points from 99.45% to 97.01% with the new unseen class. On the other hand, the classification accuracy for QVAE-BayesRadar is affected only by 1.8% point, with a classification accuracy of 98.08% in comparison to an original accuracy of 99.9%. This demonstrates the advantage of the proposed BayesRadar over LSTM-based temporal smoothing of feature embedding vectors for target classification. The plot also helps to understand correlations between estimated latent embedding for the unseen class and pretrained target classes. Since the unseen target class is generated from classes 1, 3, and 4, most false identification is seen for class 3 and 4, which corresponds to the cyclist target group. Due to strong reflection from cyclists in comparison to pedestrians, the confusion for class 1 is almost close to zero.

In parallel to t-SNE plot, similar to Table 8.3, clustering scores for the unseen class is also calculated to evaluate generalization for the proposed BayesRadar.

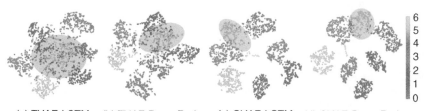

  (a) TVAE-LSTM   (b) TVAE-BayesRadar   (c) QVAE-LSTM   (d) QVAE-BayesRadar

Figure 8.15 t-SNE plot over the latent embedding for unseen data using TVAE- and QVAE-based integrated framework. Where (a) and (c) represents LSTM framework, (b) and (d) corresponds to BayesRadar [8].

Table 8.4 Quantitative analysis on the quality of the clustering over unseen classes estimated by Encoder-LSTM and BayesRadar [8].

Clustering metric	TVAE-LSTM	TVAE-BayesRadar	QVAE-LSTM	QVAE-BayesRadar
Silhouette score	0.33	**0.36**	0.59	**0.70**
Davies–Bouldin score	1.12	**1.06**	0.63	**0.56**

Table 8.4 shows the advantage of BayesRadar over LSTM with minimal reduction in both silhouette and Davies–Bouldin scores. While silhouette score for QVAE-based BayesRadar and LSTM are affected minimally, TVAE-based framework shows poor performance by reduction of $\approx$ 12% point. Similar to the t-SNE plot and clustering score over new unseen class, confidence over the unseen target class is evaluated using estimated variance from pretrained BFCN and encoder for BayesRadar and LSTM, respectively.

Class uncertainty Figure 8.16 shows the highest uncertainty over the unseen class by the LSTM in comparison to the BayesRadar. In addition, the variance estimated by BayesRadar on the unseen target class (6) has similar uncertainty score across

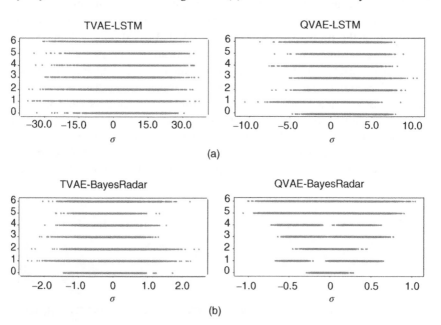

Figure 8.16 (a) A visual understanding on the uncertainty estimates over unseen classes from encoder-LSTM in comparison with (b) BayesRadar [8].

cycle target group classes (3, 4, 5) for both QVAE and TVAE. This helps to understand the spatial similarity between new target class and cycle target classes. This also shows that the sampled new class information is biased by cycles target class micro-Doppler signatures.

8.5 Summary

The demonstrated BayesRadar, a framework, wherein the embedding vectors from a variational encoder is fed into a Kalman filter for temporal smoothening and classification. This is obtained by following the tracked embedding vector from the Kalman filter. It is demonstrated that both state-of-art approach and BayesRadar achieve similar classification performance on pretrained target classes; however, BayesRadar shows much better performance against new unseen target classes. LSTM being a data-driven network suffers from rather inferior performance in case of unseen data, while the BayesRadar generalizes and performs well under such scenarios. Furthermore, the uncertainty for both seen and unseen classes for BayesRadar are much lower compared to their LSTM counterparts.

Due to inherent nature of Bayesian framework in learning model $\mathcal{M}$ from sampled data $\mathcal{D}$, it is not only used for quantification of uncertainty. It also provides flexibility and mathematical formulation for applications such automatic relevance determination (ARD), data compression, and model pruning. The ARD helps to select import input or hidden feature inside model by taking width θ_σ of model parameter which corresponds to model weight of each of the input unit or hidden unit. The θ_σ indirectly represent the precision of the features, i.e., inverse θ_σ of the prior. If θ_σ goes to infinity (very large value), precision goes to infinity, i.e., relevance goes to zero, i.e., weights for those features are ignored. On the other hand, if θ_σ is a small and finite number, feature coming out of the parameter is relevant. The similar principle can be used for pruning the weight and data compression by extending it to all hidden units up to input layer. The framework also gives flexibility for optimization of θ_σ, known as type two maximum likelihood.

8.6 Questions to the Reader

- What are the limitations of point-estimate deep learning architecture?
- How does the Bayesian paradigm affect the representation learning and what are the different kinds of uncertainties which Bayesian framework can quantify.
- What are the limitations of Bayesian framework and at what condition Bayesian and point-estimate deep neural network are same?
- What are the limitations of Laplacian and MCMC Bayesian approximation techniques?

- Imagine you are implementing a deep neural network architecture: which uncertainty you would model in which scenario?
- Given what you read in this chapter, can Bayesian framework in practice be applied, with good results, also to recursive problems? Explain why.

References

1 Kendall, A. and Gal, Y. (2017). What uncertainties do we need in Bayesian deep learning for computer vision? *Proceedings of the 31st International Conference on Neural Information Processing Systems*, ser.NIPS'17. Red Hook, NY, USA: Curran Associates Inc., pp. 5580–5590.

2 Jerrum, M. (2021). Fundamentals of partial rejection sampling. *CoRR*, vol. abs/2106.07744. https://arxiv.org/abs/2106.07744.

3 Blei, D.M., Kucukelbir, A., and McAuliffe, J.D. (2017). Variational inference: a review for statisticians. *Journal of the American Statistical Association* 112 (518): 859–877. http://dx.doi.org/10.1080/01621459.2017.1285773.

4 Robert, C. and Casella, G. (2011). A short history of Markov chain Monte Carlo: subjective recollections from incomplete data. *Statistical Science* 26 (1): http://dx.doi.org/10.1214/10-STS351.

5 Tanaka, T. (1998). A theory of mean field approximation. *Proceedings of the 11th International Conference on Neural Information Processing Systems*, ser. NIPS'98. Cambridge, MA, USA: MIT Press, pp. 351–357.

6 Kullback, S. and Leibler, R.A. (1951). On information and sufficiency. *Annals of Mathematical Statistics* 22: 79–86.

7 Dubey, A., Santra, A., Fuchs, J. et al. (2021). A Bayesian framework for integrated deep metric learning and tracking of vulnerable road users using automotive radars. *IEEE Access* 9: 68-758–68-777.

8 Dubey, A., Santra, A., Fuchs, J. et al. (2021). Bayesradar-Bayesian metric-Kalman filter learning for improved and reliable radar target classification. 2021 IEEE 31st International Workshop on Machine Learning for Signal Processing, pp. 1–6.

9 Zheng, L., Yang, Y., and Hauptmann, A. (2016). Person re-identification: past, present and future. *ArXiv*, vol. abs/1610.02984.

10 Ishfaq, H., Hoogi, A., and Rubin, D. (2010). TVAE: Triplet-based variational autoencoder using metric learning.

11 Lin, A. and Ling, H. (2006). Three-dimensional tracking of humans using very low-complexity radar. *Electronics Letters* 42 (18): 1062–1063.

12 Kim, Y. and Ling, H. (2009). Through-wall human tracking with multiple Doppler sensors using an artificial neural network. *IEEE Transactions on Antennas and Propagation* 57 (7): 2116–2122.

9

Geometric Deep Learning

At the end of this chapter, reader will have an understanding on

- An overview on the principle of learning theory behind geometric deep learning architecture.
- Different elemental blocks required to formulate geometric deep learning and different architectures.
- Applications of Bayesian geometric learning and Graph Neural Networks with their advantages for non-Euclidean datasets.

As mentioned in Chapter 8, the word "deep learning" refers to the method of learning highly nonlinear and complex problems by building them from structured and hierarchical data. The success of such a model has broad range of problems such as computer vision, natural language processing, and audio/speech analysis. The success on these tasks relies on the inherent geometry of data which are either Euclidean or grid-like.

For almost many centuries, the word "geometry" was synonymous with Euclidean geometry simply because no other types of geometries existed. In the nineteenth century, when Lobachevsky, Bolyai, Gauss, Riemann, and others constructed the first examples of non-Euclidean geometries together with the development of projective geometry, the Euclid's monopoly came to an end and entire set of different geometries emerged. Later in year 1872, mathematician Felix Klein redefines geometry as the study of invariant or symmetries of the properties that remain unchanged under some class of transformations. This approach created clarity by showing that different geometries could be defined by an appropriate choice of symmetry transformations formalized using the language of group theory [1]. As a result, different neural network architectures for different types of data need to be revisited to understand the relations between different methods and unifying principle of geometric learning.

Methods and Techniques in Deep Learning: Advancements in mmWave Radar Solutions, First Edition.
Avik Santra, Souvik Hazra, Lorenzo Servadei, Thomas Stadelmayer, Michael Stephan, and Anand Dubey.
© 2023 The Institute of Electrical and Electronics Engineers, Inc. Published 2023 by John Wiley & Sons, Inc.

The term "geometric deep learning" serves two purposes: first, to provide a common mathematical framework to derive the most successful neural network architectures such as CNN, RNN, transformers, and second, to give a constructive procedure to build future architectures in a principled way. Thus, it can be said that geometric deep learning is an umbrella term used to generalize current structured deep neural network models over non-Euclidean data and domain such as graphs and manifolds [2].

In this chapter, first, we give a theoretical understanding and a mathematical formulation of geometric deep learning using graph neural networks (GNNs). Later, we walk through different concepts of GNN based on architectural design. At the end, in order to show how these formulations and architecture can be utilized in real-life problems, two practical applications are demonstrated using automotive radar. For this purpose, radar point-cloud representation is used, and the state-of-the-art graph-based geometric deep learning is utilized in combination with Bayesian Deep Learning.

9.1 Representation Learning in Graph Neural Network

In a simplest setting, deep learning is essentially a function estimation problem where the outputs of the function over seen training datasets are known. With the availability of high-quality large datasets and compute power, the possibility to approximate any continuous function to any desired accuracy is favorable. This property is known as universal approximation which is limited to the problem definition in low dimensions [3]. The curse of dimensionality is always making such a naive learning approach impossible. In case of high-dimensional images, intuitively they have a lot of structure that can be broken and thrown away when we parse the image into a vector or follow similarity with their neighbors which helps to enable local receptive filters, known as convolution filters [4]. The typical CNN architecture encodes the locality pixels that are close to each other and are probably going to be far more strongly related than pixels that are far away. Thus, CNN encodes concept of translation in-variance which means that a pattern is interesting no matter where it is localized in the input dimension. This concept is very useful for images which follow Euclidean geometry.[1] In contrast to this, graphs don't assume any particular geometry to their nodes.

9.1.1 Fundamentals

At the first glance, the learning approach used by CNN over high-dimensional input data show two fundamental principles of translation: invariance and concept

---

1 It is important to note that, CNN architecture can be treated as a part of GNNs where each pixel is formulated as nodes of a graph.

of locality. This works very good for Euclidean data having fixed geometry, but for non-Euclidean geometry, it might be challenging to use CNN architectures, although these two fundamental principles draw cues which help in designing and formulating the GNNs.

Permutation Invariance: The first one comes from the geometry of the input signal and is known as a geometric prior. For simplicity, considering the case of image classification where the input image is not just a d-dimensional vector, but it is defined in a (most commonly) two-dimensional grid, and thus the structure of domain is captured by a symmetry. Additionally, the underlying convolution filter operations are manifested through what is called the group representation which is simply the shift operator (known as strides) of matrix (known as convolution kernel or filter) that acts on the group of neighboring points [5]. This operation makes CNN translation shift in-variance and equivariance. A similar concept is required for GNNs where nodes are volatile and thus network should be in-variance and equivariance to permutation of graph. That means if you have two graphs that are completely isomorphic,[2] the network will output exactly the same result for those two graphs. This brings to us the need of permutation in-variance formulation of the graph.

Permutation Equivariance: The second one is known as multiscale priors, which comes from the multihierarchy of domain signal created by under or over sampling nearby points. This operation can be related by coarse graining operator P applied over input scales. Another objective of CNN $f(.)$ optimization is evaluated if it is locally stable across different scales of input representation. These two principles give us a general blueprint of geometric deep learning that we can recognize in the majority of popular deep neural architectures. Typically, a sequence of equivariant layers and then an invariant global pooling layer is aggregated into a single output; in some cases, a hierarchy of coarsening procedure that takes the form of local pooling in neural network are implemented. This is a very general design that can be applied to different types of geometric neural network (Figure 9.1).

9.1.2 Learning Theory

In a typical setting of GNN, architecture has nodes and edges connecting these nodes together. For a simplicity on the formulation of GNNs, let us assume a graph with no edge and have a set of nodes v where each one of these nodes may have a feature vector x_i for node i. The typical format used for graph representation

---

2 In a graph theory, an isomorphism of graphs G and H is an edge-preserving bijection between the vertex sets of G and H such that any two vertices u and v of G are adjacent in G if and only if isomorphism over u and v are adjacent in H.

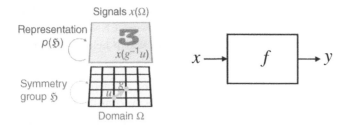

Figure 9.1 An illustration of geometric priors: the input signal (image $x \in \mathcal{X}(\Omega)$) is defined on the domain (grid Ω), whose symmetry (translation group (5) acts in the signal space through the group representation $\rho(g)$ (shift operator). Assuming on how the functions f (e.g., image classifier) interacts with the group restricts the hypothesis class [6].

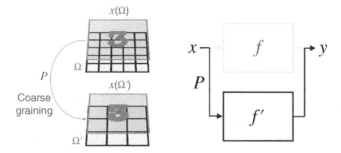

Figure 9.2 An illustration of scale separation, where we can approximate a fine-level function f as the composition $f \approx' \circ P$ of a coarse-level function f' and a coarse-graining operator P [6].

learning is by stacking these row vectors into a matrix of shape $n \times k$ where n represents nodes with k dimension feature. The biggest advantage of such a representation is that it provides order in which network will traverse. Taking the fundamental requirement of GNN into account, i.e., permutation in-variance, the output of GNN $f()$ should not depend on the alignment of nodes inside a matrix such that the function applied over such matrix give the same result, i.e.,

$$f(\mathbf{PX}) = f(\mathbf{X}) = Pf(\mathbf{X}) \tag{9.1}$$

Here, P is a mathematics operation known as permutation matrices for n nodes. The most common permutation agnostic functions ψ are summation, averaging, and maximization (Figure 9.2).

Typically, the dimension of feature vector for nodes is of the very dimension, and thus commonly it is projected into latent dimension $h_i = \psi(x_i)$. This operation is performed by considering an independent function ψ applied over row of feature vector x_i for a specific node. Later, permutation agnostic aggregation function ψ is performed over latent feature vector. The most common permutation agnostic functions ψ are summation, averaging, and maximization. This flow gives typical learning of any GNN architecture and can be formulated as follows:

$$f(\mathbf{X}) = \phi\left(\bigoplus_{i \in \mathcal{V}} \psi(\mathbf{x}_i)\right) \tag{9.2}$$

Here, $\oplus$ represent permutation agnostic aggregation function. Both ϕ and ψ are learnable functions where ϕ update the feature of operated node and ψ transforms the feature set of neighbors [7].

The GNN $f(\mathbf{X}) = \phi(\oplus_{i \in \mathcal{V}} \psi(\mathbf{x}_i))$ is now further formulated for graphs having edges such that a set of edges e which is a subset of the Cartesian product of the nodes. Typically, these edges with an adjacency matrix A is represented as follows:

$$a_{ij} = \begin{cases} 1 & (i,j) \in \mathcal{E} \\ 0 & \text{otherwise} \end{cases} \tag{9.3}$$

These edges are commonly represented in the form of adjacency matrix a which has one entry wherever there is an edge and zero, otherwise. There are other additions one can add to an adjacency matrix such as edge features, edge types, and so on, but it is ignored deliberately to keep the formulation simple. The main goal which we want to evaluate is that if main desiderata of permutation in-variant and equivariance still hold, i.e.,

$$\text{Invariance}: \quad f\left(\mathbf{PX}, \mathbf{PAP}^\top\right) = f(\mathbf{X}, \mathbf{A})$$

$$\text{Equivariance}: f\left(\mathbf{PX}, \mathbf{PAP}^\top\right) = \mathbf{P}f(\mathbf{X}, \mathbf{A})$$

This just gives the updated versions of the invariance and equivariance rules for graph with node-only inputs. The main difference is that the permutation applied to any nodes means that same permutation is applied on the edges because the rows of the node feature matrix correspond to both the rows and the columns of the adjacency matrix. The permutation matrix is conveniently written down as $\mathbf{P}$.

The concept of in-variance and equivariance is applied to nodes considering them statistically independent. However, as mentioned earlier, the notion of local functions that operate only over a locality of a certain node is very interesting for GNNs too. This enables graphs to give a very nice context to identify the node's neighborhood. For example for a node, a graph can define its one hop neighborhood as all of the nodes that are adjacent to it there exists an edge connecting them. Thus, for a node i, its (1-hop) neighborhood is commonly defined as follows:

$$\mathcal{N}_i = \{j : (i,j) \in \mathcal{E} \vee (j,i) \in \mathcal{E}\} \tag{9.4}$$

It is important to note that, explicitly directed edges are not considered, and often, we assume $i \in N_i$. Accordingly, this gives possibility to have local set of features from the neighborhood too, denoted as follows:

$$\mathbf{X}_{\mathcal{N}_i} = \left\{\{\mathbf{x}_j : j \in \mathcal{N}_i\}\right\} \tag{9.5}$$

Here, x_j denotes all the feature vectors of the nodes adjacent to node i. Considering a local function $g(.)$ which can operate over a node and its neighborhood giving multiset of features. This gives path to a GNN for generic learning by enabling generic permutation equivariance function that can be used appropriately in a shared manner by operating over this multiset of features from node and neighborhoods: $g\left(\mathbf{x}_i'\mathbf{X}_{Ni}\right)$.

In order to ensure that using locality is still equivariant, $g(.)$ needs to make sure that it does not depend on the order in which neighborhood is presented. So $g(.)$ should be typically constructed to be a permutation invariant to the entries of $\mathbf{x}_i$. As a result, GNN can be expressed as a local shared function $g(.)$ applied to every node and its neighborhood in isolation:

$$f(\mathbf{X}, \mathbf{A}) = \begin{bmatrix} - & g\left(\mathbf{x}_1, \mathbf{X}_{\mathcal{N}_1}\right) & - \\ - & g\left(\mathbf{x}_2, \mathbf{X}_{\mathcal{N}_2}\right) & - \\ & \vdots & \\ - & g\left(\mathbf{x}_n, \mathbf{X}_{\mathcal{N}_n}\right) & - \end{bmatrix} \tag{9.6}$$

The visual understanding of this can be expressed in Figure 9.3. Here, local shared function $g(.)$ is applied to every node and its neighborhood in isolation, and as a result of this node, we will be transformed from input features x_i to latent features h_i. Later, the transformed feature matrix can be used for a task like node classification $z_i = f(h_i)$, where one can look at each of the nodes latent in isolation and learn a classifier on them. Graph classification $z_G = f(\oplus_{j\in\mathcal{N}_i}h_i)$, where permutation invariant function such as summation can be used to combine all the nodes into one representation and then apply graph-level classifier to classify them. At last, link prediction $z_{ij} = f(h_i, h_j, e_{ij})$, where certain properties of edges can predicted using a function that operates over the latency of the two nodes and potentially any edge features between them. These are the general tasks for GNNs.

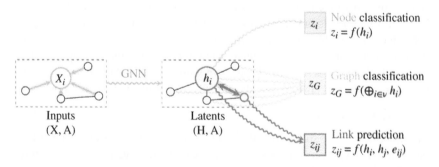

Figure 9.3 An illustration of feature engineering and different task learning for GNNs based on node, edges, or entire graph.

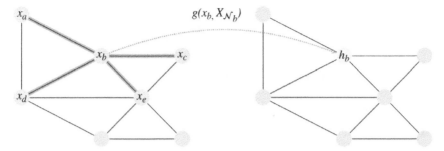

Figure 9.4 An illustration of efficient task-independent feature learning, i.e., embedding learning where nodes are encoded into the latent so that similarity in the embedding space (e.g., dot product) approximates similarity in the original network.

9.1.3 Embedding Learning

During the processing embedding learning, the feature of the nodes of the graph is embedded into the embedding space where every node has a set of s feature coordinates. The goal during embedding learning is to find the embedding so that the similarity of the node in the embedding space relates or is similar to the both root and neighboring node from the original network. This is done by using an encoder which maps node to embeddings, as illustrated in Figure 9.4. This further brings the requirement to define a similarity function, i.e., a measure of similarity in the original network, i.e., $(x_b, x_c) \approx \mathbf{h}_{x_b}^{\mathsf{T}} \mathbf{h}_{x_c}$. The most common way to measure similarity in literature is using the Euclidean distance between the two points, whereas this gets computationally expensive and inefficient for high-dimensional non-Euclidean data. As an alternative, cosine similarity is often used. The cosine similarity is defined as follows: if all feature sets are nonnegative, dot product between two vector equals the cosine of the angle between the two vectors. If the two vectors are orthogonal, then the similarity is zero, and if they overlap one on top of the other, then it is basically the length of the vector. In GNNs, the cosine similarity is just the dot product between the coordinates node x_b and neighboring nodes. Thus, during embedding learning, parameter of encoder is optimized using defined similarity function.

9.2 Graph Representation Learning

As mentioned earlier, GNN constructs permutation equivalent functions $f(.)$ by sharing a local permutation invariant function $g(.)$[3] over all the neighborhoods.

---

3 Under many studies, $g(.)$ is referred to as diffusion propagation or a message passing.

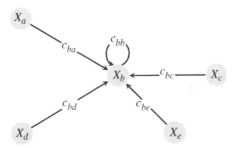

Figure 9.5 An illustration of convolution CNN where features of neighbors are aggregated with fixed weights [6].

Broadly, choice and alignment of ψ, ϕ, and $\oplus$ give three flavors of GNN layers, i.e., convolution, attention, and message passing. The details are discussed below.

9.2.1 Convolution GNN

In the convolution setting of GNNs, features from the neighbors are aggregated with constant weight c_{ij} which depends directly on the adjacency matrix, as illustrated in Figure 9.5. This tells us how much does a node x_i value the features of node x_j and then they are basically coefficients in a well weighted sum or a different kind of weighted combination of locally transformed features using ψ function applied to every node feature in isolation. The permutation invariant aggregators such as summation can be used to return the recipe for the features of that node in the next step. The operation can be formalized as follows:

$$\mathbf{h}_i = \phi\left(\mathbf{x}_i, \bigoplus_{j \in \mathcal{N}_i} c_{ij} \psi\left(\mathbf{x}_j\right)\right) \tag{9.7}$$

The most common examples are Chebyshev networks [8], graph convolution networks [9], and the simplified graph convolution networks [10]. They are very useful when edge encoded labels are similar. This can be solved with a very simple average of what is inside the node of a graph. Additionally, because of their scalability, lightweight, easy to train, and hardware friendly with simple matrix operation, such an architecture is very useful:

9.2.2 Recurrent Graph Neural Networks (RGNNs)

Recurrent graph neural network (RGNN) [11] processes the node information recurrently by assuming that the nodes exchange information with their neighbors until a stable point is reached. RGNN defines the node aggregation function as in Eq. (9.8).

$$\mathbf{h}_u^t = \sum_{v \in N(u)} \phi(\mathbf{x}_u, \mathbf{x}_{(u,v)}^e, \mathbf{x}_v, \mathbf{h}_v^{(t-1)}) \tag{9.8}$$

where $\phi(\cdot)$ is a nonlinear differentiable recurrent function. In [11], the proposed architecture is an RNN, where the connections between the neurons are classified into internal and external connections. While the first refers to the internal connections within units of the network, the external ones refer to the edges of the processed graph.

9.2.3 Graph Autoencoders (GAEs)

Graph autoencoders (GAEs) belong to the family of unsupervised frameworks and are used for graph-based representation learning and graph generation [12]. In both tasks, an encoder is employed to extract node embeddings of a graph, followed by the reconstruction of new graphs from corresponding latent or embedding vectors. For representation learning, graph structural information is reconstructed as an adjacency matrix. In the case of graph generation, the process might involve a stepwise generation of the nodes and edges, or output the entire graph at once.

9.2.4 Spatial Temporal Graph Neural Networks (STGNNs)

Spatial temporal graph neural networks (STGNNs) aim at capturing underlying spatial and temporal relations simultaneously [12]. The spatial relation is captured by using graph convolutions, and the temporal relation is modeled by employing RNN blocks.

9.2.5 Attention GNN

On the other hand, when edges are no longer just as a code for label similarity but may contain some repulsion effects from the neighboring node, constant weights are not enough to encode such effect. For example if x_i retweet someone's tweet, it doesn't mean that x_i entirely agrees it. As a result, constant fixed weights are replaced with learnable scalar weights. It is done by attention mechanism which is any learnable function $a(.)$ that takes features of the root node and neighbor nodes and returns a coefficient that can be used to weight the node's contributions to the root node. The entire process can be formulated as follows:

$$\mathbf{h}_i = \phi\left(\mathbf{x}_i, \bigoplus_{j \in \mathcal{N}_i} a\left(\mathbf{x}_i, \mathbf{x}_j\right) \psi\left(\mathbf{x}_j\right)\right) \tag{9.9}$$

This is a very elegant and powerful way to learn more complicated weighted combinations, but it still doesn't require that much information has be computed and stored, and there are several models implementing this idea, and some of the

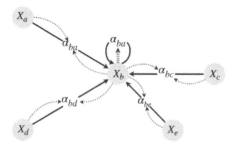

Figure 9.6 An illustration of attention CNN where features of neighbors are aggregated with implicit weights (via attention) [6].

earliest ones include Monet model [13], the graph attention network [14], and the gated attention network [15]. These models are very useful as a sort of middle ground for situation where edges do not encode strict homophily relations and one might want some more complicated sums with only one scalar for every edge. The typical illustration of such architecture is summarized in Figure 9.6, where features of neighbors are aggregated via implicit weight learning using attention.

9.2.6 Message-passing GNN

Unlike attention GNNs where only raw feature of neighbors is aggregated via implicit weight learning using attention, message-passing GNN have both root and neighboring node features collaborate to compute an arbitrary vector using ψ function, as formulated below.

$$\mathbf{h}_i = \phi\left(\mathbf{x}_i, \bigoplus_{j \in \mathcal{N}_i} a\left(\mathbf{x}_i, \mathbf{x}_j\right) \psi\left(\mathbf{x}_j\right)\right) \tag{9.10}$$

This formulation gives the most generic GNN layer and can fit some very complex simulation data such as algorithmic reasoning, physical simulations, or computational chemistry, but that could imply some scalability or learnability issues, because now these weights needs to be stored and computed on an entire vector for every edge in your graph. Typically, in a graph, there are a lot more edges than there are nodes. The typical message passing GNN is illustrated in Figure 9.7

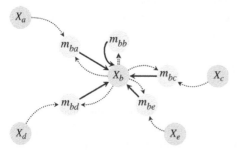

Figure 9.7 An illustration of attention message-passing GNN where arbitrary vectors are computed using feature set of both neighbors and nodes [6].

9.3 Applications

As discussed above, Geometric Deep Learning techniques can be particularly useful to capture underlying local and global structures while working with representations that involve use of nongrid points. One such representation is point cloud, which is composed of multiple points in a N-dimensional space usually having the coordinate values as features and in some cases, additional domain-specific features such as Doppler value of the point for a radar use-case. In this section, we cover two of our recent works [16] with radar point clouds as input in both the applications. Application one involves using a novel Bayesian PointNet architecture that allows for an end-to-end target classification and localization. The second application leverages cross-learning idea from Chapter 5 and GNNs discussed in this chapter earlier to build a robust gesture-sensing solution for long-range (target distance from radar > 1 m).

9.3.1 Application 1: Long-Range Gesture Recognition

Human control interfaces for various indoor applications, e.g., sound systems, television, and lightning, continuously evolve to increase user comfort. Going from buttons, switches, and rotary knobs to touchscreen-based control with smartphone-apps, the trend is now heading toward gesture-based control [17, 18], allowing for interaction from further distances without requiring additional hardware. Gesture control needs to be intuitive, accurate, and should be computationally cheap [19]. While vision-based systems nowadays fulfill these criteria, they are often seen as too privacy-invasive. Therefore, radar-based systems have recently been under investigation as a privacy-preserving alternative solution [20–24]. However, while progress has been made, and radar-based gesture sensing has been deployed in products [20, 25–27], it is still less accurate than vision-based systems, especially on longer distances. Therefore, we present an mmWave radar-based gesture-recognition solution that incorporates knowledge from a camera system during the training process. Our proposed system uses preprocessed radar data in the form of point clouds, as input to a DGCNN [28] for performing cross-learning from a camera point cloud, which then predicts the performed gesture, taking multiple frames into account.

9.3.1.1 Camera Point Cloud

To determine 3D joint coordinates from 2D camera images, the joints must be visible in at least two cameras and must be matched. First, we feed all camera images into Detectron2 [29] to detect people and their associated keypoints in 2D space. Here we use the best pretrained keypoint recurrent convolution neural network (RCNN) architecture available in the Detectron2 model zoo trained on the Coco keypoint dataset [30]. The Coco keypoint dataset contains 17 keypoints for each person, which are the nose, left eye, right eye, left ear, right ear, left shoulder, right shoulder, left elbow, right elbow, left wrist, right wrist, left hip, right hip, left knee, right knee, left ankle, and right ankle. Matching 2D poses through multiple views is challenging for various reasons, such as occlusion

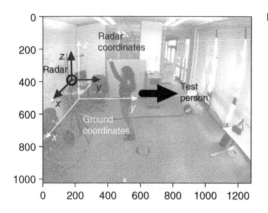

Figure 9.8 Radar setup: $\theta_{tilt} = 0$.

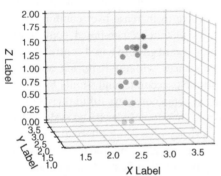

Figure 9.9 Camera point cloud.

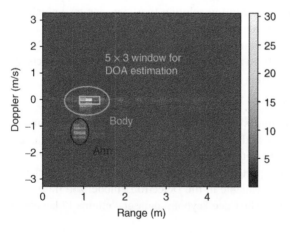

Figure 9.10 Range-Doppler image. Shaded bar shows the intensity of each bin.

Figure 9.11 Radar point cloud with shade showing velocity value in m/s.

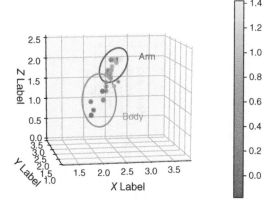

and truncation (Figures 9.8–9.11). Our association component, as shown in Figure 9.12, consists of two robust approaches that complement and reinforce each other: person re-identification and epipolar geometry. The result of our approach is as shown in Figure 9.9, 3D joint coordinates, and 17 keypoints.

Person Re-identification The goal of person re-identification is to identify the corresponding person of interest in different views or at different times. We use the implementation provided in [31]. It uses a ML model, in which the embeddings of individuals that are close to each other are considered to be the same person.

Epipolar Geometry Using epipolar geometry, the geometric information can be leveraged and objects localized in different cameras. For example, if the object's location is specified in the first camera, the search area in the second camera is constrained to a single line if the epipolar geometry is known [32].

9.3.1.2 Radar Point Cloud
This section describes the radar sensor setup and the preprocessing for converting raw radar data to five-dimensional point clouds [33] with x–y–z coordinates, intensity, and Doppler values, as illustrated in Figure 9.13.

9.3.1.3 mmWave FCMW Radar Sensor
We use Infineon's BGT60TR13C FMCW radar chipset. It operates by transmitting multiple frames, each containing a sequence of frequency chirps with a short ramp time. The response is digitized in 12-bit by the analog-digital converter, and is further passed to the PC over USB. The operating parameters of the radar are presented in Table 9.1.

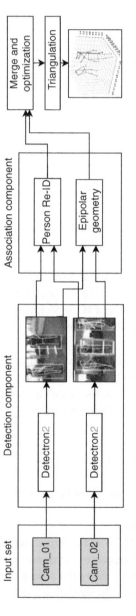

Figure 9.12 Overview of used approach for skeleton detection. Given an input set of calibrated camera recordings from different views, we first detect persons and their keypoints in 2D space using Detectron2. Thus, to match 2D poses in multiple views, we combine person re-identification and geometric information of persons. The person re-identification cues along with the geometric cues are then concatenated and triangularized to determine 3D points.

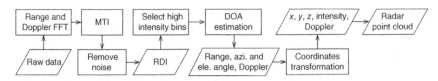

Figure 9.13 Preprocessing of raw radar data.

Table 9.1 Operating parameters.

Parameters, symbol	Value
Ramp start frequency, f_{min}	57.5 GHz
Ramp stop frequency, f_{max}	58.5 GHz
Frame rate, f	20 fps
Number of samples per chirp, NTS	64
Sampling frequency, fs	2 MHz
Chirp time, T_c	64 μs
Number of chirps, PN	128
Number of Tx antennas, N_{Tx}	1
Number of Rx antennas, N_{Rx}	3

Range-Doppler and Angles Estimation The reflected target signal is mixed with the transmitted chirp signal and then passed through a low-pass filter to obtain the intermediate frequency signal [34]. Since we are only interested in moving targets in gesture recognition, we use previous frame subtraction as MTI [35]. By applying 2D Range-Doppler FFT along the sample and chirp axes, we can acquire range-Doppler image, comprising range, Doppler, and intensity information, as shown in Figure 9.10. Additionally, as we work with single-person scenarios, we first detect the range-Doppler bin with the highest intensity and filter out any points far away from it as noise [36]. Afterward, we select a certain number of high-intensity range-Doppler bins and use the values in a 5 × 3 window around each bin, seen in Figure 9.10, to estimate the covariance matrix for DOA estimation with bartlett beamforming. Now, the points are in spherical radar coordinates $(r, \theta_{azi}, \theta_{ele})$, which correspond to range, azimuth angle, and elevation angle. We finally apply the transformation matrix [37] shown below,

$$\begin{bmatrix} x \\ y \\ z \end{bmatrix} = \begin{bmatrix} 1 & 0 & 0 \\ 0 & \cos\theta_{tilt} & \sin\theta_{tilt} \\ 0 & -\sin\theta_{tilt} & \cos\theta_{tilt} \end{bmatrix} \begin{bmatrix} r\cos\theta_{ele}\sin\theta_{azi} \\ r\cos\theta_{ele}\cos\theta_{azi} \\ r\sin\theta_{ele} \end{bmatrix} + \begin{bmatrix} x_r \\ y_r \\ h \end{bmatrix} \tag{9.11}$$

where (x, y, z) denote the resulting ground Cartesian coordinates, θ_{tilt} and (x_r, y_r, h) are the tilt angle and ground Cartesian coordinates of radar, respectively.

Point Cloud With the help of a radar sensor, we can collect the points corresponding to the detected object. However, a specific number of sample points for each frame is required as input. Thus, to simplify the computation, traditional zero-padding is the method we choose for oversampling, which means adding all-zero points to complement the point cloud. Meanwhile, for frames with more points than we require, sorting the points by their intensity and selecting a predefined number allows us to save the samples with the most information [38]. Finally, we take (x, y, z, d) intensities, which are coordinates, Doppler, and intensity value into consideration as the features of each point, shown in Figure 9.11.

9.3.1.4 Architecture and Learning

In this section, we introduce the network's architecture of our proposed autoencoder shown in Figure 9.14 for cross-learning between radar and camera information [39]. The architecture is composed of a GNN [40], specifically DGCNN [28], followed by LSTM layers for recognition of gesture sequences.

Edge Convolution The radar point cloud contains n points with F features in each frame, denoted by $\mathbf{R} = \{\mathbf{r}_1, \dots, \mathbf{r}_n\} \subseteq \mathbb{R}^F$. In our case, $F = 5$, representing 3D coordinates x_i, y_i, z_i, intensity and Doppler value i, d in RD-image and n points per frame. We apply k-nearest neighbor (k-NN) by Euclidean distance to generate the local graph of $\mathbf{R}$ in $\mathbb{R}^F$ including self-loop, represented as $\mathcal{G} = (\mathcal{V}, \mathcal{E})$, where $\mathcal{V} = \{1, \dots, n\}$ and $\mathcal{E} \subseteq \mathcal{V} \times \mathcal{V}$ are the *vertices* and *edges*, respectively. As illustrated in Figure 9.15, to capture both global structure and local neighborhood information, *edgefeatures* between two points is computed as follows:

$$e_{ij} = h_{\Theta}(\mathbf{r}_i, \mathbf{r}_j - \mathbf{r}_i) \tag{9.12}$$

where $h_{\Theta} : \mathbb{R}^F \times \mathbb{R}^F \to \mathbb{R}^{F'}$ is a nonlinear function with learnable parameters Θ and $\{\mathbf{r}_j : (i, j) \in \mathcal{E}\}$ is the set of neighbors around $\mathbf{r}_i$. More specifically, this edge convolution can be expressed as follows:

$$e_{ijf'} = \sigma(\theta_{f'} \cdot (\mathbf{r}_j - \mathbf{r}_i) + \phi_{f'} \cdot \mathbf{r}_i) \tag{9.13}$$

and implemented as a shared MLP, here $\Theta = (\theta_1, \dots, \theta_{F'}, \phi_1, \dots, \phi_{F'})$, *leakyReLU* is chosen as nonlinear function $\sigma(\cdot)$. At last, we take max pooling as the aggregation operation on the *edgefeatures* to update the points:

$$r'_{if'} = \max_{j:(i,j) \in \mathcal{E}} e_{ijf'} \tag{9.14}$$

which can capture the sharpest features to represent the points in lower-level. In general, this EdgeConv creates an F'-dimensional point cloud with the same number of points as the input F-dimensional point cloud, where $k = 3$ and $n = 64$.

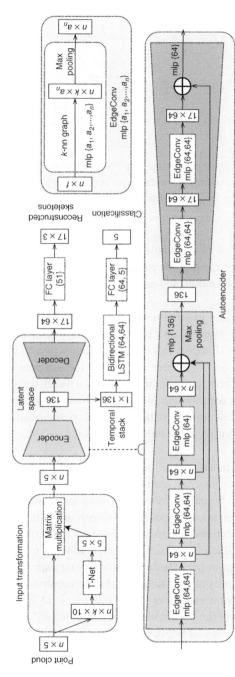

Figure 9.14 Network architecture: Radar point cloud in shape $n \times 5$ first passes through an input transformation module and then is fed into DGCNN autoencoder, followed by one fully connected layer to reconstruct camera-based skeletons. Next, the latent space of each autoencoder-frame is stacked in temporal order with sequence length $l = 30$, which is used to generate classification scores for five classes by two bidirectional LSTM layers and two fully connected layers. EdgeConv block: A tensor of shape $n \times f$ is fed into the block as input, for computing edge features of each point in k-nn Graph by a MLP with the number of layer units denoted as $\{a_1, a_2, \ldots, a_n\}$. Then, after max pooling across neighboring edge features, we get a updated tensor of shape $n \times a_n$.

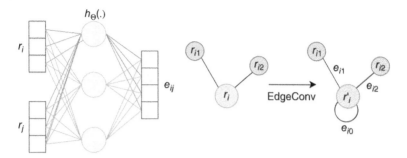

Figure 9.15 Left: Generating f'-dimensional edge features $\mathbf{e}_{ij}$ from two f-dimensional points $\mathbf{r}_i$ and $\mathbf{r}_j$. Here, we choose one fully connected layer as $h_\Theta(\cdot)$ and $f = 3, f' = 4$, for example. Right: Updating $\mathbf{r}'_i$ from $\mathbf{r}_i$ by Edge Convolution, $k = 3$ in our case, and e_{i0} indicates self-loop.

Model Architecture The overview of our whole network architecture is visualized in Figure 9.14, including three main parts: input transformation, cross learning between radar and camera point cloud, and gesture recognition.

Input Transformation As described in Section 9.3.1.2, we multiply the coordinates in radar coordination by our radar setup's rotation matrix to convert them to ground Cartesian coordinates. However, this transformation is insufficiently precise due to the measurement, and the object, in this case the test person, should be invariant following certain geometric modifications. As a result, we refer to the spatial transformation in this contribution [41], in which a mininetwork (T-net in Figure 9.14) predicts an affine transformation matrix from the point itself and its neighbors, and applies it directly to the input point cloud.

Cross Learning The upper branch of the network architecture illustrated in Figure 9.14 is the autoencoder for cross learning between the point cloud from radar and camera domain. The encoder contains three successive EdgeConv blocks, where the graph is dynamically updated after each one. Afterward, their multiscale outputs are concatenated together, followed by one fully connected layer to form 136 dimensional latent space. Then, the decoder is used to reconstruct camera skeletons from prior latent feature vector, having a similar structure with the encoder, while there are only two EdgeConv blocks present. Finally, after going through the last fully connected layer (51), the point cloud is reshaped to 17×3, representing 17 joints of the person in (x, y, z) coordinates. The number k for computing k-NNs is set as 3 for all EdgeConv blocks.

Classification After obtaining the latent features from the encoder, the lower branch shown in Figure 9.14 extracts both spatial and temporal information of each gesture motion sequence with length $l = 30$. The latent feature vectors of each frame in the sequence are first stacked to create $l \times 136$ tensor, fed as input into two 64-units bidirectional LSTM layers [42]. The last two fully connected layer (64,5) produce five-dimensional feature vector for gesture classification.

9.3.1.5 Experiments and Results

We used a synchronized camera–radar setup as introduced in Section 9.3.2.1 to perform seamless multimodal data acquisition to train our proposed model and then for evaluation. The training data are balanced and consist of 1773 recordings and 392 recordings (sequence length $l = 30$ for each recording) for evaluation. The evaluation dataset doesn't include any camera data as only radar data are used as input during inference.

We define five macrogestures as classes that involve a complete hand movement. The five classes are swipe – where the hand is moved horizontally from one end to another; push – where the hand is moved toward the radar from the body; pull – where the hand is moved from away from the radar toward the body; clockwise – where a big circular movement is done by the hand in a clockwise manner; and anticlockwise – where the same is done but in an anticlockwise manner. The choice of gestures was made based on their relevance to past literature and simplicity. The training data involved five volunteers who performed the gesture at a distance ranging between 1 and 2 m with minimal prior instructions and switched between left and right hands and their distance from the radar. The evaluation dataset was similarly performed by another five volunteers but in a different room and radar orientation setup to demonstrate the proposed architecture's generalization capability. Furthermore, the room setup for the evaluation dataset contained more reflecting objects so as to check our model's ability to suppress such reflections.

The upper branch in Figure 9.14 is first trained with synchronized radar point cloud and camera skeletons data using MSE loss, and then we save the **input transformation** and **Encoder** modules for the training of classification in the lower branch by cross-entropy and triplet losses combined in the same weight, which can be set as frozen or not for fine-tuning. The overall accuracy for the proposed model is 98.4% compared to from 45% to 90% for the baseline model, which is trained in a unimodal fashion. Figure 9.16 depicts the confusion matrix for our proposed model and Figure 9.17 represents the training and testing accuracy of different setups. It clearly demonstrates the superiority of the proposed learning approach over a unimodal approach with better accuracy, robustness, and stable learning.

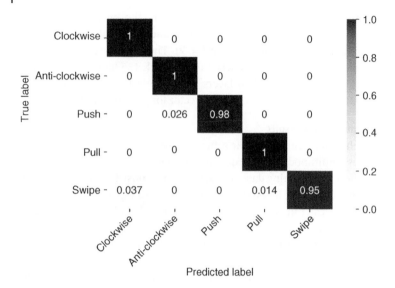

Figure 9.16 Confusion matrix.

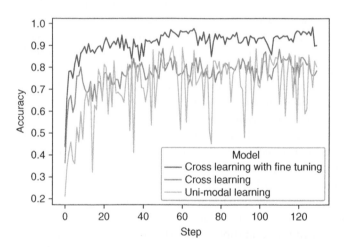

Figure 9.17 Accuracy comparison among unimodal, cross learning, and cross learning with fine-tuning.

9.3.2 Application 2: Bayesian Anchor-Free Target Detection

The typical GNN architecture brings an instance of geometric diplomatic blueprint with the permutation group as the geometric prior. A sequence of permutation equivalent layers, often referred to as propagation or diffusion layers, and an optional global pooling layer, to produce a single graph-wise output, are used by the architecture. Additionally, some architectures also include local pooling layers that can also be learnable. In principle, a graph with no edges is a set which are also unordered. In this case, the most straightforward approach is to process each element in the set entirely independently by applying a shared function to their feature vectors. This translates into a permutation equivalent function over the set. This is a special setting of GNN architecture known as DeepSets in deep learning or PointNet [43] in computer graphics. Taking this concept and architecture inspired from PointNet, we proposed a framework for target classification using radar point cloud. The architecture details are below.

The given point cloud is denoted as $\mathbf{P}$ which contains n points $p_1, p_2, \ldots, p_n \in \mathbb{R}^d$ with d dimensional features. The input feature vector of each point p_i for segPointNet consists of the global target coordinate space (x_i, y_i), the azimuth angle (θ), and the signal reflection power (σ). The set of semantic labels is denoted as $\mathbf{L}$. Semantic segmentation of a point cloud is a function Ψ which assigns semantic labels to each point in the point cloud, i.e., $\Psi : \mathbf{P} \mapsto L^n$. The objective of segmentation algorithms is to find the optimal mapping from the input space to the semantic labels. However, the performance of the network strongly depends on the richness of input features presented to the network [44].

In this work, we connect the concepts of multimodality and attention to split the problem of target detection into three parts, as illustrated in Figure 9.18.

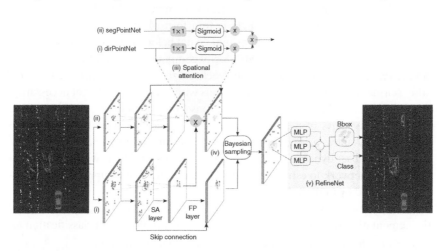

Figure 9.18 Overview of our methods which is composed of five modules, (i) target's direction vector estimation (dirPointNet); (ii) hierarchical spatial attention inside target segmentation network (segPointNet) using target's direction modality; (iii) hierarchical spatial attention module; (iv) Bayesian sampling for target location using direction vector; and (v) refineNet for bounding box (bbox) and target class estimation. Both dirPointNet and segPointNet follow the same architecture parameter with sampling abstraction (SA) layer and feature propagation (FP) layer [16].

First, a one-channel direction field vector is estimated for each point. This outputs a coherent direction vector for all points belonging to an unique target. Afterward, a direction field vector is used to provide attention inside the segmentation network to achieve better feature learning. In the end, the information from the segmented output and the direction field network is fused to perform a Gaussian sampling for unique cluster identification. These clusters are passed through another network for a bounding box estimation. The individual networks uses the PointNet++ [43] architecture adapted for the radar scene data [45]. We propose an end-to-end hierarchical, spatial, attention-based multimodal segmentation framework, called HARadNet. The first network, named dirPointNet, learns the direction vector field for each point. This information is used inside the second network (segPointNet) to improve spatial localization. This approach increases the cross-correlation of the shared representations and potentially yields a faster convergence.

9.3.2.1 Direction Field Estimation

Using the inherent property of the radar sensor, the tangential velocity of the target is determined along with its spatial position in the azimuth dimension (θ) for each reflected point. Taking advantage of both values together with the sensors' yaw angle (ϕ), the direction of motion for each point is estimated as $d(\varphi) = \theta + \phi$. The velocity of the target is compensated with respect to the ego-motion, while the azimuth angle is transformed to global coordinates. In real-world scenarios, many reflections that do not belong to a moving object show a nonzero ego-motion-compensated velocity component, caused by errors in the odometry, sensor misalignment, time synchronization errors, mirror effects, or other sensor artifacts. In addition, reflections with zero velocity do not necessarily belong to a static object, since also reflections from the bottom of a rotating car wheel or body parts of a pedestrian that move perpendicular to the walking direction may show no radial velocity. As a result, multiple static targets are misinterpreted as dynamic ones. To overcome this problem, we optimize the dirPointNet network to estimate the direction of targets and suppress unwanted "noise" caused by multipath reflections.

The network follows an encode–decode scheme similar to a general semantic segmentation network, except for changing the problem from classification to regression. The network is trained using 4D input feature tensors ($\hat{x}_{cc}, \hat{y}_{cc}, \hat{\theta}_{cc}, \vec{v}_r$) to predict the motion direction vector for each traffic participant ($d(\varphi)$). Furthermore, both input features and labels are rescaled to the range of $[0,1)$ by applying

$$\hat{x}_{cc} = \frac{x_{cc} - x_{i_{cell}}}{s_{x_{cell}}}, \hat{y}_{cc} = \frac{y_{cc} - y_{k_{cell}}}{s_{y_{cell}}}, \vec{v}_r = \frac{1}{v_{max}}\vec{v}_r, \hat{\theta}_{cc} = \frac{\theta}{60°}, d(\hat{\varphi})_i = \frac{d(\varphi)_i}{180°}$$

$$(9.15)$$

with $(x_{i_{\text{cell}}}, y_{k_{\text{cell}}})$ representing the position of the left bottom corner of a cell. The indices (i, k), $(s_{x_{\text{cell}}}, s_{y_{\text{cell}}})$ resemble the cell extension in x, y-direction, while $v_{\max}$ is the maximum velocity, and $\sigma_{\max}$ the maximum signal power obtained from the whole data set. This rescaling restricts the gradient from exploding during the network training.

The resulting network optimization is still very challenging due to highly unbalanced target points and the sparsity of target features. Thus, we propose the following hybrid loss function to train the network.

$$
\begin{aligned}
L_{\text{direction}} &= w_{\text{wmse}} L_{\text{wmse}} + (1 - w_{\text{wmse}}) L_1 \\
L_{\text{wmse}} &= \frac{1}{n} \frac{\sum_{i=0}^{n} w_i \left(\hat{d}(\varphi)_i - d(\varphi)_i \right)^2}{\sum_{i=0}^{n} w_i} \\
w_i &= \log \left(\hat{d}(\varphi)_i + 1 \right) + 1
\end{aligned}
\tag{9.16}
$$

Here, L_1 represents absolute differences between the true value and the predicted value. The value of w_i is calculated over a number of positive samples in a batch. An empirically determined fixed value of 0.8 as used for weighted mean square error (w_{wmse}).

9.3.2.2 Direction Attention

Due to sparsity, nonuniformity, and the highly imbalanced nature of target representations in radar point clouds, the actual target recognition becomes very challenging. Here we use pointwise multiplication to provide hard spatial attention inside the segPointNet. As dirPointNet and segPointNet share the same number of input tensors, we are able to preserve the flexibility of providing attention at different feature abstraction levels of the network. segPointNet uses $(\hat{x}_{\text{cc}}, \hat{y}_{\text{cc}}, \vec{v}_{\text{r}}, \hat{\sigma})$ as input feature tensor, where $\hat{\sigma}$ is the normalized signal reflection power. We have used pointwise multiplication to provide hard attention inside network. For this purpose, prior to attention, both estimated segPointNet and dirPointNet outputs are standardized between 0 and 1. Additionally, 1×1 is a pointwise convolution with a residual connection to attention that is used to avoid vanishing gradient. Due to the difference in input features, the estimated signal from both segPointNet and dirPointNet is standardized to range between 0 and 1 using a sigmoid function prior to the spatial attention inside segPointNet. Both network are optimized using end-to-end training. As a result, they complement each other in learning target features from different modalities. The total loss is formulated as follows:

$$
\begin{aligned}
L_{\text{attention}} &= w_{\text{cls}} L_{\text{cls}} + (1 - w_{\text{cls}}) L_{\text{direction}} \\
L_{\text{cls}} &= -\left(1 - \hat{p}_y \right)^{\gamma} \log \left(\hat{p}_y \right)
\end{aligned}
\tag{9.17}
$$

Table 9.2 A comparison of the effect of attention at different feature abstraction layers for both binary (top-row) and multiclass (bottom-row) segmentation using the F1-score (see Eq. (9.18)).

w/o Attention				1-depth Attention				2-depth Attention				3-depth Attention			
Avg	Ped	Car	Bike	Avg	Ped	Car	Bike	Avg	Ped	Car	Bike	Avg	Ped	Car	Bike
0.92	—	—	—	0.95	—	—	—	**0.97**	—	—	—	0.96	—	—	—
0.88	0.61	0.85	0.90	0.92	0.43	0.97	0.77	0.94	0.46	0.98	0.64	**0.95**	**0.75**	**0.98**	**0.91**

The bold values signify highest accuracy for binary and multi-class segmentation with different attention-based training setup.

where L_{cls} denotes the loss for the point classification from the scene and $L_{direction}$ is for the direction field estimation of classified radar targets. In Eq. (9.17), $y \in \{0, \dots, K-1\}$ represents an integer class label, $\hat{\mathbf{p}} = (\hat{p}_0, \dots, \hat{p}_{K-1}) \in [0,1]^K$ is a vector representing the estimated probability distribution over the K classes and γ is a focusing parameter which specifies how much high-confidence predictions contribute to the overall loss. Table 9.2 gives an insight on the effect of spatial attention by dirPointNet on segPointNet by evaluating the network performance using the F1-score. The F1-score is the harmonic mean between precision P and recall R, given by

$$F_1 = \frac{2 \cdot P \cdot R}{P + R} \tag{9.18}$$

The performance of segPointNet is evaluated for both binary and multi-class segmentation tasks. While the binary segPointNet is optimized to predict only the foreground as targets of interest and the background as static-reflections or noise, multiclass segPointNet is trained to preserve the target class and background. The binary segPointNet demonstrates a better average F1-score of 0.92 in contrast to the multiclass segPointNet with an average F1-score of 0.88. This is caused by the unavailability of uniform features between points of the same class. In contrast to the case without attention, both binary and multiclass segPointNet show an improvement in the average F1-score with attention applied inside the network and validates advantage and scalability of our proposed architecture. Further, the multiclass segPointNet shows a significant improvement in the average F1-score being equal to 0.95 when attention is applied at every feature propagation layer. It is comparable to binary segPointNet with an average F1-score of 0.96. Additionally, the network shows major improvements in the recognition of pedestrians, which share the least samples of target distribution in the dataset [45]. This proves the advantage of attention inside a network with different modalities which acts as an additional target feature point and improves the scene recognition.

Table 9.3 Distribution of the six different target class in the dataset.

Car (%)	Pedestrian (%)	Pedestrian group (%)	Bike (%)	Truck (%)	Static (%)
1.23	0.31	0.74	0.11	0.60	97.01

Source: Adapted from Schumann et al. [45].

9.3.2.3 Experiments

The evaluation of the proposed framework is done on real-world data that was collected by four automotive corner radar sensors. All reflections that belong to the same physical object are grouped and annotated with a corresponding label from the following classes: car, pedestrian, pedestrian group, bike, truck, and static. The distribution of the occurrences among the six classes is shown in Table 9.3. This gives a clear indication of the typical foreground vs. background class imbalance, present in the data. Furthermore, pedestrians and bikes have the least number of training samples. Additionally, both share a lower signal reflection strength and sparse point distribution in comparison to the other target classes. As a result, the segPointNet struggles to categorize these classes correctly, as shown in Table 9.2. In addition, Table 9.3 shows the distribution of object availability with respect to the distances to the ego-vehicle. Following the object distribution over distance, we process cropped scenes within a range of 80 m for $\hat{x}_{cc}$ and an absolute range of 20 m for $\hat{y}_{cc}$. This reduces the total number of static targets during the network training.

9.3.2.4 Bayesian Sampling

Compared to other vision tasks such as segmentation or categorization, localizing the object is a very complex task, mainly because the same region could also jointly belong to another target, if it is closely located or partially occluded. Additionally, due to the in-homogeneous and sparse distribution of radar points in the point cloud space, points from neighboring regions often have similar characteristics. Therefore, we intend to seek a method that has the ability to discover potential and meaningful patterns among proximity points, so that the set of points can be clustered into unique groups in a robust way.

In order to deal with such situations, our attention direction network can be guided not only toward the more relevant features but also toward the selection of unique regions, using a cluster of direction vectors as a signature distribution, in combination with spatial information. For ease of usage, we call this step Bayesian sampling, which is performed in two steps. At first, both spatial and direction information is fused and passed to different algorithms to estimate the possible number (order) of targets available in the scene. This is used by the clustering algorithm to find and localize them uniquely. In our experiment, we used the density-based spatial clustering of applications with noise (DBSCAN)

Table 9.4 Comparison of the class-agnostic clustering methods for target localization with normalized features, i.e., compensated spatial location in $\hat{x}_{cc}$, $\hat{y}_{cc}$ and azimuth $\hat{\theta}$), compensated target velocity ($\hat{v}_r$), reflected signal strength $\hat{\sigma}$, and motion direction vector ($\widehat{d(\varphi)}$) estimated from dirPointNet [9].

Normalized feature dimensions						Clustering score (meanIoU [%])			
$\hat{x}_{cc}$	$\hat{y}_{cc}$	$\vec{v}_r$	$\hat{\sigma}$	$\hat{\theta}$	$\widehat{d(\varphi)}$	BIC + GM	DBSCAN + GMM	DBSCAN + VBGM (Dirichlet process)	DBSCAN + VBGM (Dirichlet distribution)
✓	✓					0.195	0.700	0.703	0.711
✓	✓	✓				0.731	0.640	0.747	0.746
✓	✓	✓	✓			0.328	0.722	0.739	0.734
✓	✓	✓	✓	✓		0.497	0.593	0.601	0.598
✓	✓	✓	✓	✓	✓	0.640	0.652	0.661	0.652
✓	✓	✓	✓		✓	0.445	0.796	0.806	0.808
✓	✓	✓			✓	0.481	0.850	0.859	0.858
		✓			✓	**0.815**	**0.894**	**0.893**	**0.894**

The bold values signify highest clustering score using combination of different set of normalized features and methods.

and the Bayesian information criterion (BIC) for the estimation of target order. Thereafter, the output is clustered into the desired unique bins using a Gaussian mixture model (GMM) operated at different feature dimensions. To the best of the author's knowledge, this approach is novel and has not been investigated in literature before.

While Table 9.2 demonstrates the quantitative advantage of direction vector attention for radar point-target segmentation, Table 9.4 shows the advantage of the direction vector as the key feature dimension for target clustering. To compare the performance of localization, clustering algorithms are evaluated for different dimensions of features ($\hat{x}_{cc}$, $\hat{y}_{cc}$, $\vec{v}_r$, $\hat{\sigma}$, $\hat{\theta}$, $\widehat{d(\varphi)}$). Further, to evaluate the generalization of our approach, the performance of clustering is evaluated for different clustering algorithms, i.e., GMM and its variant, variational Bayesian Gaussian mixture (VBGM) model. The principle behind VBGM is the same as for GMM, which is expectation minimization, but VBGM also adds a regularization by integrating prior distribution information. Although priors may bring initial biases, the VBGM selects a suitable number of effective clusters (targets) by avoiding the singularities which are often found in expectation-maximization solutions, and pushing weights values close to zero.

$$\text{IoU} = \frac{\left| b_{gt} \cap b_{pred} \right|}{\left| b_{gt} \cup b_{pred} \right|} \tag{9.19}$$

A common evaluation metric for segmentation tasks is the mean intersection over union (mIoU). The IoU compares a predicted bounding box b_{pred} with the ground truth bounding box b_{gt} and is defined as Eq. (9.19). Here, $| \cdot |$ measures the total area of the underlying set. If the ground truth and the predicted bounding box are almost identical, the IoU score tends to be close to one. If the two bounding boxes do not overlap, the IoU score will be zero. While $\widehat{d(\varphi)}$ helps to increase the localization accuracy, due to nonuniformity in spatial dimensions $(\hat{x}_{\text{cc}}, \hat{y}_{\text{cc}}, \hat{\sigma})$ of the radar-point clouds, the performance of clustering is strongly limited. As a result, only the estimated direction vector $(\widehat{d(\varphi)})$ and its magnitude $(\vec{v}_r)$ are considered for localization. This leads to significant improvement in localization accuracy with a mean IoU of $\approx 90\%$ using DBSCAN and GMM and its variant, validating the robustness of each feature combination. Figure 9.19 illustrates the statistics of IoU for all clustering methods over the entire validation scene for all targets of interest. A large percentage of all scenes is distributed around 1.0 in the vertical axis of IoU for all four subplots, indicating that many scenes in the sequence achieve a perfect match. The combination of DBSCAN and GMM

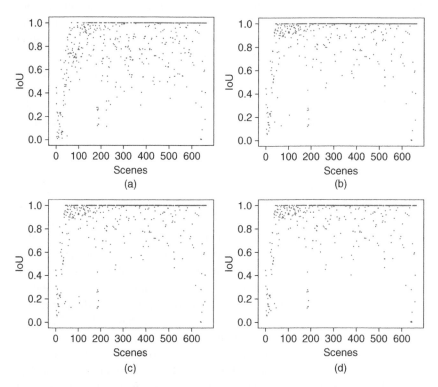

Figure 9.19 Visualization of the IoU population distribution for all target of interest over the entire test sequence. (a) IoU with BIC + GMM. (b) IoU with DBSCAN + GMM. (c) IoU with DBSCAN + VBGM1. (d) IoU with DBSCAN + VGGM2.

gains over 50%, and finally the percentage of scenarios with an mIoU over 0.8 exceeds 82%. The best performance is achieved by the combinations of DBSCAN with VBGM using a Dirichlet distribution as priors. As a result, the rest of our experiments and evaluations are based on this clustering method.

Figure 9.20 illustrates the different stages for target localization using the proposed Bayesian sampling and the effect of the estimated direction vector from the dirPointNet. To have a deeper insight on the advantages of the approach, multiple examples together with their behavior on boundary conditions are demonstrated using four random scenes from the validation set. The top row shows the input point cloud marked with the ground truth label and the direction vector. The middle two rows show the output of the segPointNet and the dirPointNet. The last row follows the output of the Bayesian sampling layer. Figure 9.20a shows multiple false positives from both segPointNet and dirPointNet. The Bayesian sampling layer, however, suppresses false positives by fusion and later discards all the points with high variance in their feature dimension due to the high variability of the direction vectors between the neighboring points. As a result, the network successfully finds regions of interest. Figure 9.20c shows multiple detections for the same target and misdetections for the target due to noncoherent direction vectors. Multiple detections for the same target can be suppressed by nonmaxima suppression (NMS) in the postprocessing stage. The examples demonstrate that the performance of Bayesian sampling strongly relies on the $\overline{d(\theta_{cc})}$ feature in comparison to the distributed and nonuniform spatial feature dimension $(\hat{x}_{cc}, \hat{y}_{cc})$. Furthermore, Figure 9.20b demonstrates an interesting observation and advantage of our approach where both segPointNet and dirPointNet predict a target. The Bayesian sampling layer, however, discards both points due to noncoherent direction vectors and spatial sparsity (no neighborhood). Thus, our approach also helps to suppress false positive detections.

9.3.2.5 Multi-task Learning

After a joint end-to-end learning of multiclass segmentation and direction field estimation using a input dimension of $4 \times n$, the region of interest in the form of unique point clusters, with the dimension of $4 \times m$, is passed through a refineNet, conceptually similar to [46]. The refineNet is a multihead network with a regression and classification head. While regression head predicts parameters of a 2D bounding box (Bbox), i.e., its center (x_c, y_c) and its size (l, w) around the clustered points, the classification head is optimized to predict the target class for the points. For the box center estimation, a residual-based 2D localization is performed, similar to [47], where the network estimates the centroid over the center. To guarantee a fixed number of input points to the FC layer, the sampling process during training is considered. For the 2D bounding box estimation, up to 32 points are randomly sampled from the point clusters for every radar target.

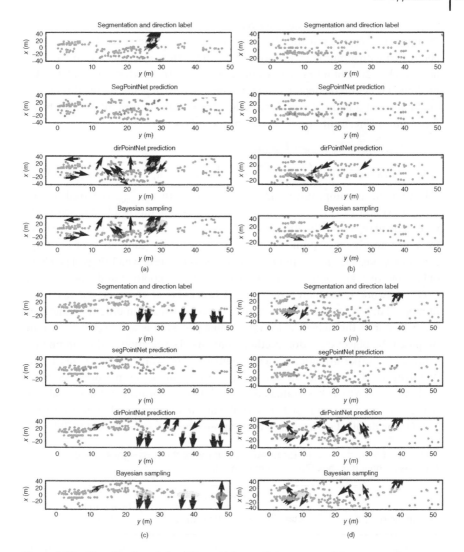

Figure 9.20 Visualization of the different intermediate outputs of the proposed a three-stage approach. The first row in each of (a–d) represents the spatial input with the corresponding label and its associated direction field. The last three rows of (a–d) show the predicted segmentation, estimated direction vector, and Bayesian sampling for the region of interest estimation [9].

The labeling process automatically generates a bounding box from the annotated radar point targets by using the ground truth as a reference. The training is performed with a multitask loss for joint optimization of segmentation, direction field, and a 2D bounding box estimation. Since the performance of the box prediction relies on the region proposal which in turn depends on the dirPointNet prediction, a trained network is used for the initialization of all weights before the training. The, multitask loss is defined as follows:

$$L_{\text{multi-task}} = L_{\text{attention}} + L_{\text{refineNet}}$$

$$L_{\text{refineNet}} = w_{\text{Bbox}} \left(L_{c_x,\text{reg}} + L_{c_y,\text{reg}} + L_{h,\text{ reg}} + L_{w,\text{ reg}} \right) + w_{\text{s, cls}} L_{\text{s, cls}} \qquad (9.20)$$

During the training, the weight for a target with $L_{\text{attention}}$ is handled using Eqs. (9.16) and (9.17). $L_{c_x,\text{reg}}$ and $L_{c_y,\text{reg}}$ are used for the residual-based center regression of the box estimation network. Furthermore, $L_{w,\text{ reg}}$ and $L_{h,\text{ reg}}$ are losses for width and height estimation, while $L_{\text{s, cls}}$ is for the estimation of the target class of the box with w_{Bbox} and $w_{\text{s,cls}}$ their respective task weightings. The choice of loss for box regression and target classification is smoothL1 and cross-entropy.

The network is evaluated on the full test data set to cover the maximum number of different situations, including the corner cases, in order to understand the behavior of the networks for targets like pedestrians and cyclists. Since the proposed 2D object detection method contains classification and bounding box estimation, the performance of these modules will be evaluated using the F1-score (compare Eq. (9.18)) and the IoU metric. Additionally, the performance of the multihead box network is evaluated under different training conditions and compared with both region based state-of-the-art architecture and regression-based YOLO. The detailed performance of all approaches are summarized in Table 9.5.

Table 9.5 Comparison of the localization accuracy for the class-agnostic and class-aware bounding box (Bbox) estimation [9].

	Task weights		Classification	Bbox
Weighting criteria	Class ($w_{\text{s,cls}}$)	Bbox (w_{Bbox})	(F1-score)	(mIoU)
Class only	1	0	0.92	0.89
Bbox only	0	1	**0.95**	0.86
Empirical weighting	0.5	2	0.78	0.93
SOTA [48]	—	—	0.64	0.64
YOLO	2	7,2,5,0.5	0.66	0.32
Weights with task uncertainty [49]	—	—	0.82	**0.96**

The bold values signify highest classification and location (detection) accuracy for proposed framework when trained in different setup and compared to SOTA and YOLO architecture.

First, the performance of the network is evaluated separately for cluster classification and class-agnostic bounding box (Bbox) estimation. This is done by enabling either $w_{s,cls}$ or w_{Bbox} exclusively. The mIoU for the class-only scenario is the same as the mIoU for the target localization since the Bbox is not fine-tuned for clustered points but achieves a classification accuracy of 0.92. On the other hand with Bbox-only, the target localization is degraded from mIoU of 0.89 to 0.86, while the classification accuracy in this case is the same as the F1-score from the multiclass segmentation task. Both centroid and corner points are estimated without knowing the target class (point distribution), thus the network considers all points as an inlier and tries to fit the bounding box to it, which results in a drift of the centroid. This leads to a bad localization accuracy of 0.86 for the multiclass segmentation network. By learning the target class distribution together with the bounding box estimation, the network shows a slightly better localization accuracy with an IoU of 0.93 and a classification accuracy of 0.78. The strong weighting for w_{Bbox} in contrast to $w_{s,cls}$ results from the much stronger average classification loss, compared to the box regression.

Additionally, the proposed framework is compared with a state-of-the-art region-based object detection [48] and a regression-based detection algorithm using YOLO architecture [50]. The SOTA network shows the best target classification of 0.96, at the cost of localization mIoU of 0.64. The improved target classification score is due to the reason that the performance of SOTA is evaluated over accumulated multiple frames over 500 ms, in contrast to our approach where the network is evaluated for a single frame. Additionally, the feature dimension used for localization refinement and classification of clustered point includes the original input dimensions ($\hat{x}_{cc}$, $\hat{y}_{cc}$, $\hat{\theta}$, $\hat{\sigma}$) and not the estimated direction vector ($\vec{v}_r$). As a result, the network struggles to classify clustered point-clouds due to incoherency between spatial features. Further, we also compared our proposed framework with the YOLO architecture by optimizing it directly on our normalized input data over a single frame. The YOLO results in worse localization and classification accuracy of 0.32 and 0.66, respectively.

Although the multitask approach aims to improve the learning efficiency by learning multiple objectives from shared representations, the performance of the multitask network optimization is highly sensitive to weights (w_{Bbox}, $w_{s,cls}$) given to the different losses (L_{bbox}, L_{cls}). In contrast to an expensive grid-search or naive weighted sum of losses, the network is optimized using online learned weights with task (homoscedastic aleatoric) uncertainty which captures relative confidence between tasks, motivated by Cipolla et al. [49] and Bischke et al. [51]. As a result, the Bbox and target class estimation is modified with joint-learning function $p\left(\mathbf{y}_1, \mathbf{y}_2 \mid \mathbf{f}^{\mathbf{W}}(\mathbf{x})\right)$, where $\mathbf{y}_1$ and $\mathbf{y}_2$ represents box regression and target classification as two outputs from the multihead network $\mathbf{f}^{\mathbf{W}}(\mathbf{x})$. This leads to the minimization objective of $-\log p\left(\mathbf{y}_1, \mathbf{y}_2 \mid \mathbf{f}^{\mathbf{W}}(\mathbf{x})\right)$ for the multioutput model, given as follows:

$$
\begin{aligned}
&= -\log p\left(\mathbf{y}_1, \mathbf{y}_2 = c \mid \mathbf{f}^{\mathbf{W}}(\mathbf{x})\right) \\
&= -\log \mathcal{N}\left(\mathbf{y}_1; \mathbf{f}^{\mathbf{W}}(\mathbf{x}), \sigma_{y_1}^2\right) \cdot \text{Softmax}\left(\mathbf{y}_2 = c; \mathbf{f}^{\mathbf{W}}(\mathbf{x}), \sigma_{y_2}\right) \\
&= \frac{1}{2\sigma_{y_1}^2}\left\|\mathbf{y}_1 - \mathbf{f}^{\mathbf{W}}(\mathbf{x})\right\|^2 + \log \sigma_{y_1} - \log p\left(\mathbf{y}_2 = c \mid \mathbf{f}^{\mathbf{W}}(\mathbf{x}), \sigma_{y_2}\right) \\
&= \frac{1}{2\sigma_{y_1}^2}\mathcal{L}_1(\mathbf{W}) + \frac{1}{\sigma_2^2}\mathcal{L}_2(\mathbf{W}) + \log \sigma_{y_1} \qquad\qquad (9.21) \\
&\quad + \log \frac{\sum_{c'} \exp\left(\frac{1}{\sigma_{y_2}^2} f_{c'}^{\mathbf{W}}(\mathbf{x})\right)}{\left(\sum_{c'} \exp\left(f_{c'}^{\mathbf{W}}(\mathbf{x})\right)\right)^{\frac{1}{\sigma_{y_2}^2}}} \\
&\approx \frac{1}{2\sigma_{y_1}^2}\mathcal{L}_1(\mathbf{W}) + \frac{1}{\sigma_{y_2}^2}\mathcal{L}_2(\mathbf{W}) + \log \sigma_{y_1} + \log \sigma_{y_2}
\end{aligned}
$$

Here, $\mathcal{L}_1(\mathbf{W})$ stands for the *smoothL*$_1(\mathbf{y}_1, \mathbf{f}^{\mathbf{W}}(\mathbf{x}))$ for the regression loss $\mathbf{y}_1$ and $\mathcal{L}_2(\mathbf{W}) = -\log \text{Softmax}\left(\mathbf{y}_2, \mathbf{f}^{\mathbf{W}}(\mathbf{x})\right)$ for the cross-entropy loss $\mathbf{y}_2$. The network is trained to predict the log-variance $\log \sigma_y^2$ for more numerical stability and avoids gradient division when the loss is zero. The network shows the best localization accuracy, compared to our proposed framework and SOTA. This is due to the reason that the localization loss is very sensitive to both the estimated corner and the center points. As a result, the task uncertainty-based approach helps the network to choose an appropriate loss weighting during the training. While this approach helps the network to improve the classification accuracy by significant amounts compared to the empirical weighting, the network still struggles to classify very sparse and spatially distributed clustered points into the desired target class without using the estimated direction vector as another feature dimension.

In addition to the quantitative evaluation, Figure 9.21 illustrates few corner cases of our proposed anchor-free detection framework and its localization and classification accuracy. Figure 9.21a1 shows multiple overlapping box proposals around the ground truth target due to varying target distribution. Consequently, the concept of NMS can be used as postprocessing. Thus, all the boxes having IoU > 0.5 with the ground truth are considered. Figure 9.21a2 illustrates the updated bounding box tightly coupled with the ground truth. Similarly, both Figure 9.21d1,d2 demonstrate the effectiveness of HARadNet for successful localization of a target with just four points. Additionally, it also preserves the target class. As a result, the need for predefined anchor boxes or grid-based regression methods can be eliminated. On the other hand, while Figure 9.21b,c demonstrates the target localization, the estimated bounding box leaves out some target points treating as an outlier. As a result, the network contributes to the

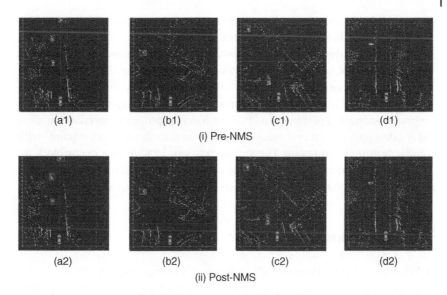

Figure 9.21 Combined illustration of the performance of HARadNet. (i) shows the class-aware bounding box estimation, and (ii) illustrates the final prediction after using nonmaxima suppression (NMS) as postprocessing [9].

false-negatives during the target localization. This is caused by the loss function (L_{box}) which does not penalize loss caused by the background and the foreground separately. As an alternative, in the future, loss functions similar to the one proposed in [52], can be used for better localization.

9.4 Conclusion

In this chapter, we have covered ideas starting from understanding non-Euclidean data structures, the challenges they come with and geometric deep learning architectures to tackle them. This would enable the readers to develop a solution around non-Euclidean radar data representations instead of relying only on specific representation and conventional deep learning architectures such as CNNs. The applications provide more in-depth view on how to use such ideas in real-life radar use-cases. In application one, the use of GNNs along with a cross-learning methodology allows for capturing target dynamics at a longer range while suppressing other targets. The can be easily extended to other use-cases involving people sensing such as activity recognition or people tracking. The second application propose an anchor-free model for target localization and classification employing hierarchical spatial attention captured in the form of motion modality by using direction field vectors for each target point. This is done without using additional sensors or other dependencies such as temporal information or cross-model distillation. The entire model is trained in an end-to-end training framework using the concept of multitask learning. This approach helps the model to converge faster for the combined tasks while sharing the learned representations. The joint motion guided spatial attention mechanism for feature selection is highly essential to select useful features from the learned representations and to improve localization performance. Furthermore, the proposed Bayesian sampling layer takes both spatial and motion modality to sample and cluster points with similar feature distributions. The architecture can be used as a major block for radar use-cases ranging from vacuum cleaner bots to self-driving cars which require to be able to not only detect targets but also classify them to act accordingly.

9.5 Questions to the Reader

- What are the limitations of CNN architecture in comparison to GNNs?
- Discuss the significance of permutation invariance and permutation equivariance in a neural network architecture?
- Differentiate between attention GNNs and Message Passing GNN architectures? Which architecture would better fit for traffic forecasting and why?
- How does the geometric learning paradigm affect the graph representation learning and which two explicit conditions are applied over GNN architecture.
- In the first application, we use MSE as reconstruction loss for reconstruction of radar point cloud data to camera point cloud data. Discuss possible drawbacks of using MSE for reconstruction and propose some other loss function that might exhibit better performance.
- Given what you read in this chapter, what kind of GNN architecture is used for second application and can it be improved using concept of attention GNNs? Explain why.
- Propose a novel solution for fall detection using radar point cloud as input.

References

1 Klein, F.C. (2008). A comparative review of recent researches in geometry. https://arxiv.org/abs/0807.3161 (accessed 9 August 2022).

2 Bronstein, M.M., Bruna, J., LeCun, Y. et al. (2017). Geometric deep learning: going beyond euclidean data. *IEEE Signal Processing Magazine* 34 (4): 18–42. https://doi.org/10.1109.

3 Cybenko, G.V. (1989). Approximation by super positions of a sigmoidal function. *Mathematics of Control, Signals and Systems* 2: 303–314.

4 Fukushima, K. (2004). Neocognitron: a self-organizing neural network model for a mechanism of pattern recognition unaffected by shift in position. *Biological Cybernetics* 36: 193–202.

5 Bruna, J. and Mallat, S. (2012). Invariant scattering convolution networks. https://arxiv.org/abs/1203.1513.

6 Bronstein, M., Bruna, J., Cohen, T., and Veličković, P. (2021). Geometric deep learning: grids, groups, graphs, geodesics, and gauges. https://doi.org/10.48550/arXiv.2104.13478.

7 Zaheer, M., Kottur, S., Ravanbakhsh, S. et al. (2017). Deep sets. https://arxiv.org/abs/1703.06114.

8 Defferrard, M., Bresson, X., and Vandergheynst, P. (2016). Convolutional neural networks on graphs with fast localized spectral filtering. https://arxiv.org/abs/1606.09375.

9 Kipf, T.N. and Welling, M. (2016). Semi-supervised classification with graph convolutional networks. https://arxiv.org/abs/1609.02907.

10 Wu, F., Zhang, T., Souza, A.Hd. et al. (2019). Simplifying graph convolutional networks. https://arxiv.org/abs/1902.07153.

11 Scarselli, F., Gori, M., Tsoi, A.C. et al. (2009). The graph neural network model. *IEEE Transactions on Neural Networks* 20 (1): 61–80.

12 Wu, Z. et al. (2021). A comprehensive survey on graph neural networks. *IEEE Transactions on Neural Networks and Learning Systems* 32 (1): 4–24.

13 Burgess, C.P., Matthey, L., Watters, N. et al. (2019). MONet: Unsupervised scene decomposition and representation. https://arxiv.org/abs/1901.11390.

14 Veličković, P., Cucurull, G., Casanova, A. et al. (2017). Graph attention networks. https://arxiv.org/abs/1710.10903.

15 Zhang, J., Shi, X., Xie, J. et al. (2018). GaAN: Gated attention networks for learning on large and spatiotemporal graphs. https://arxiv.org/abs/1803.07294.

16 Dubey, A., Santra, A., Fuchs, J. et al. (2022). HARadNet: Anchor-free target detection for radar point clouds using hierarchical attention and multi-task learning. *Machine Learning with Applications* 8: 100275.

17 Wan, Q., Li, Y., Li, C., and Pal, R. (2014). Gesture recognition for smart home applications using portable radar sensors. *2014 36th Annual International Conference of the IEEE Engineering in Medicine and Biology Society*, pp. 6414–6417.

18 Smith, K.A., Csech, C., Murdoch, D., and Shaker, G. (2018). Gesture recognition using mm-Wave sensor for human–car interface. *IEEE Sensors Letters* 2 (2): 1–4.

19 Mitra, S. and Acharya, T. (2007). Gesture recognition: a survey. *IEEE Transactions on Systems, Man, and Cybernetics Part C: Applications and Reviews* 37 (3): 311–324.

20 Hayashi, E., Lien, J., Gillian, N. et al. (2021). RadarNet: Efficient gesture recognition technique utilizing a miniature radar sensor. *Proceedings of the 2021 CHI Conference on Human Factors in Computing Systems.* ACM.

21 Stephan, M., Hazra, S., Santra, A. et al. (2021). People counting solution using an FMCW radar with knowledge distillation from camera data. *2021 IEEE Sensors.* IEEE.

22 Stephan, M., Servadei, L., Arjona-Medina, J. et al. (2022). Scene-adaptive radar tracking with deep reinforcement learning. *Machine Learning with Applications* 8: 100284.

23 Servadei, L., Sun, H., Ott, J. et al. (2022). Label-aware ranked loss for robust people counting using automotive in-cabin radar. *IEEE International Conference on Acoustics, Speech and Signal Processing (ICASSP).*

24 Santra, A. and Hazra, S. (2020). Deep Learning Applications of Short Range Radars, ser. Artech House radar library. Artech House. https://books.google.de/books?id=Qb-VzQEACAAJ (accessed 9 August 2022).

25 Hazra, S. and Santra, A. (2018). Robust gesture recognition using millimetric-wave radar system. *IEEE Sensors Letters* 2 (4): 1–4.

26 Hazra, S. and Santra, A. (2019). Radar gesture recognition system in presence of interference using self-attention neural network. *2019 18th IEEE International Conference On Machine Learning And Applications (ICMLA),* pp. 1409–1414.

27 Molchanov, P., Gupta, S., Kim, K., and Pulli, K. (2015). Short-range FMCW monopulse radar for hand-gesture sensing. *2015 IEEE Radar Conference (RadarCon),* pp. 1491–1496.

28 Wang, Y., Sun, Y., Liu, Z. et al. (2019). Dynamic graph CNN for learning on point clouds. *ACM Transactions on Graphics* 38 (5): https://doi.org/10.1145/3326362.

29 Wu, Y., Kirillov, A., Massa, F. et al. (2019). Detectron2. https://github.com/facebookresearch/detectron2 (accessed 9 August 2022).

30 Lin, T.-Y., Maire, M., Belongie, S. et al. (2015). Microsoft COCO: common objects in context.

31 Luo, H., Gu, Y., Liao, X. et al. (2019). Bag of tricks and a strong baseline for deep person re-identification. *2019 IEEE/CVF Conference on Computer Vision and Pattern Recognition Workshops (CVPRW),* pp. 1487–1495.

32 Zhang, Z. (2014). *Epipolar Geometry,* 247–258. Boston, MA: Springer US. https://doi.org/10.1007/978-0-387-31439-6_128.

33 Salami, D., Hasibi, R., Palipana, S. et al. (2022). Tesla-rapture: a lightweight gesture recognition system from mmWave radar sparse point clouds. *IEEE Transactions on Mobile Computing arXiv preprint arXiv:2109.06448.*

34 Zhang, Z., Tian, Z., and Zhou, M. (2018). Latern: dynamic continuous hand gesture recognition using FMCW radar sensor. *IEEE Sensors Journal* 18 (8): 3278–3289.

35 Zhang, G., Geng, X., and Lin, Y.-J. (2021). Comprehensive mPoint: a method for 3D point cloud generation of human bodies utilizing FMCW MIMO mm-Wave radar. *Sensors* 21 (19): https://www.mdpi.com/1424-8220/21/19/6455.

36 Svenningsson, P., Fioranelli, F., and Yarovoy, A. (2021). Radar-PointGNN: Graph based object recognition for unstructured radar point-cloud data. *2021 IEEE Radar Conference (RadarConf21)*, pp. 1–6.

37 Jin, F., Sengupta, A., and Cao, S. (2022). mmFall: Fall detection using 4-D mmWave radar and a hybrid variational RNN autoencoder. *IEEE Transactions on Automation Science and Engineering* 19 (2): 1245–1257.

38 Qian, K., He, Z., and Zhang, X. (2020). 3D point cloud generation with millimeter-wave radar. *Proceedings of the ACM Interactive, Mobile, Wearable and Ubiquitous Technologies* 4 (4): https://doi.org/10.1145/3432221.

39 Wang, Y., Jiang, Z., Li, Y. et al. (2021). RODNet: A real-time radar object detection network cross-supervised by camera-radar fused object 3D localization. *IEEE Journal of Selected Topics in Signal Processing* 15 (4): 954–967. https://doi.org/10.1109.

40 Lopera, D.S., Servadei, L., Kiprit, G.N. et al. (2021). A survey of graph neural networks for electronic design automation. *2021 ACM/IEEE 3rd Workshop on Machine Learning for CAD (MLCAD)*, pp. 1–6.

41 Charles, R.Q., Su, H., Kaichun, M., and Guibas, L.J. (2017). PointNet: Deep learning on point sets for 3D classification and segmentation. *2017 IEEE Conference on Computer Vision and Pattern Recognition (CVPR)*, pp. 77–85.

42 Gong, P., Wang, C., and Zhang, L. (2021). MMPoint-GNN: Graph neural network with dynamic edges for human activity recognition through a millimeter-wave radar. *2021 International Joint Conference on Neural Networks (IJCNN)*, pp. 1–7.

43 Qi, C.R., Yi, L., Su, H., and Guibas, L.J. (2017). PointNet++: Deep hierarchical feature learning on point sets in a metric space. *CoRR*, vol. abs/1706.02413. http://arxiv.org/abs/1706.02413.

44 Schumann, O., Hahn, M., Dickmann, J., and Wöhler, C. (2018). Semantic segmentation on radar point clouds. *2018 21st International Conference on Information Fusion (FUSION)*, pp. 2179–2186.

45 Schumann, O., Hahn, M., Scheiner, N. et al. (2021). RadarScenes: A real-world radar point cloud data set for automotive applications. https://doi.org/10.5281/zenodo.4559821.

46 Qi, C.R., Su, H., Mo, K., and Guibas, L.J. (2016). PointNet: Deep learning on point sets for 3D classification and segmentation. *CoRR*, vol. abs/1612.00593. http://arxiv.org/abs/1612.00593.

47 Qi, C.R., Litany, O., He, K., and Guibas, L.J. (2019). Deep hough voting for 3D object detection in point clouds. *CoRR*, vol. abs/1904.09664. http://arxiv.org/abs/1904.09664.

48 Danzer, A., Griebel, T., Bach, M., and Dietmayer, K. (2019). 2D car detection in radar data with PointNets. *CoRR*, vol. abs/1904.08414. http://arxiv.org/abs/1904.08414.

49 Cipolla, R., Gal, Y., and Kendall, A. (2018). Multi-task learning using uncertainty to weigh losses for scene geometry and semantics. *2018 IEEE/CVF Conference on Computer Vision and Pattern Recognition*, pp. 7482–7491.

50 Simon, M., Milz, S., Amende, K., and Gross, H. (2018). Complexer-YOLO: Real-time 3D object detection on point clouds. *CoRR*, vol. abs/1803.06199. http://arxiv.org/abs/1803.06199.

51 Bischke, B., Helber, P., Folz, J. et al. (2019). Multi-task learning for segmentation of building footprints with deep neural networks. *2019 IEEE International Conference on Image Processing (ICIP)*, pp. 1480–1484.

52 Hall, D., Dayoub, F., Skinner, J. et al. (2018). Probability-based detection quality (PDQ): a probabilistic approach to detection evaluation. *CoRR*, vol. abs/1811.10800. http://arxiv.org/abs/1811.10800.

Index

Methods and Techniques in Deep Learning: Advancements in mmWave Radar Solutions, First Edition.
Avik Santra, Souvik Hazra, Lorenzo Servadei, Thomas Stadelmayer, Michael Stephan, and Anand Dubey.
© 2023 The Institute of Electrical and Electronics Engineers, Inc. Published 2023 by John Wiley & Sons, Inc.

Printed and bound by CPI Group (UK) Ltd, Croydon, CR0 4YY

23/11/2022

03165630-0002